CLIMAX in
COVENTRY

CLIMAX in COVENTRY

My life of FINE ENGINES and FAST CARS

WALTER HASSAN, OBE

in collaboration with

GRAHAM ROBSON

MERCIAN

MOTOR RACING PUBLICATIONS LTD.
Unit 6, The Pilton Estate, 46 Pitlake, Croydon, Surrey, CR0 3RY

First published in 1975
Reprinted 1997
ISBN 0 9530721 2 6

Copyright © 1975 - Walter Hassan, Graham Robson
and Motor Racing Publications Ltd.

All rights reserved. No part of this publication may be reproduced, stored in a retrieval system, or transmitted, in any form or by any means, electronic, mechanical, photocopying, recording or otherwise, without prior permission of Motor Racing Publications Ltd.

Facsimile reprint, with kind permission of the original publishers Motor Racing Publications Ltd.

Printed by Mercian Manuals Ltd
353 Kenilworth Road, Balsall Common
Coventry. CV7 7DL

Contents

Introduction		7
Chapter 1	BENTLEY BOY	10
	Bentley cars 1919-1931	25
Chapter 2	BARNATO AND BROOKLANDS	28
	Barnato-Hassan and Pacey-Hassan Specials	39
Chapter 3	RECORDS AND RADIALS	44
	Napier-Railton Land Speed Record car	55
Chapter 4	BIRTH OF THE XK120	58
	Jaguar XF to XK engine development	68
Chapter 5	A FIRE-PUMP THAT WINS RACES	71
	Coventry Climax FW single-cam engines	82
Chapter 6	NEVER BELIEVE THE OPPOSITION	85
	Coventry Climax FPE V-8 engine	99
	Coventry Climax FPF twin-cam 'four'	101
Chapter 7	THE CLIMAX FOR CLIMAX	106
	Coventry Climax FWMV V-8 engine	120
Chapter 8	THE STILLBORN SIXTEEN	123
	Coventry Climax FWMW flat-16 engine	131
Chapter 9	FINALE WITH TWELVE CYLINDERS	134
	Jaguar V-12 engine development	146
Chapter 10	AND THEY CALL IT RETIREMENT!	150
Appendix 1	Coventry Climax Grand Prix record	154
Appendix 2	Coventry Climax FMWV development record	157

To my wife and my several other bosses,
for their forbearance, faith and understanding

Introduction

When I had the pleasure, in 1971, of receiving some credit for the design of Jaguar's famous V-12 engine, it gave me quite a shock to realise that I had already been working on fine engines and fast cars for more than 50 years. Although I was already 66 years old, and past the normal retiring age, I certainly didn't feel as if I had finished with high-performance engines and work in general, and I couldn't visualise myself sitting idly in a comfortable armchair for the rest of my life.

I suppose I had thought about retirement from time to time, but as I had always been so completely involved, and usually thoroughly happy in my work, I had never really given much thought to it. I should have left my position as Chief Engineer – Power Units at Jaguar in 1970, but before the date arrived Sir William Lyons asked me to stay on for a couple of years and 'see it through' with the new engine.

It was about this time that more and more kind people seemed to want me to go and talk to them and their societies about my experiences, particularly in motor racing, and with Bentley before the last war. Also, I spent a lot of time writing a detailed technical paper (which I read in America to the Society of Automobile Engineers in 1971) about the new engine; although I seemed to have to scrap and rewrite the whole thing several times I rather enjoyed the job.

I think it was probably at that time that I decided to write a book about my life's work, but as soon as I had shed responsibilities at Browns Lane I was approached by Louis Stanley of BRM, who wanted me to give them advice and consultancy on their plans for engine developments in Grand Prix racing. Although it started with only occasional visits to Bourne, I was soon closely involved in the work which, as I write this, is beginning to bear fruit, so I can only really consider myself as semi-retired!

It began to look as if I would never have time to get everything down on paper, and I am pleased that Graham Robson finally persuaded me to tackle the task. Graham has prepared my reminiscences for publication, and has also added quite a lot of factual material about the cars and engines with which I have been linked.

I have certainly seen many changes in the art and science of engine design, and of the approach to motor racing, since I first joined Bentley Motors in 1920. The most direct comparison I can make is that those first 3-litre Bentley engines we used at Le Mans probably produced about 85 bhp, while the latest BRM 3-litre engines I have seen installed in Formula One cars develop something like 460 bhp.

I am proud of the fact that I was one of W.O.Bentley's earliest employees – in fact I was Number 14 on the clock, joining as a 15-year-old apprentice and being paid ten shillings (50p) a week – and I still recall with pleasure the many enthralling hours I spent learning about motor cars at the Cricklewood works in North-West London. It was through Bentley that I was drawn into the motor racing scene, and I have not been far away from the tracks ever since.

I suppose the period by which many people will remember my work was when Coventry Climax engines were winning races all round the world, and until my friend Keith Duckworth's DFV engine wins a few more races it is nice to know that Climax engines still hold a record number of wins (96, I believe) in Grand Prix events. Building special track-racing cars for Woolf Barnato and Bill Pacey taught me a lot in connection with the design and construction of a racing car from the tyres upward, while my relatively short stays at ERA and at Thomson & Taylor introduced me to many firm friends. As to my spell with the Bristol Aeroplane Company, I think it taught me a great deal about engineering method, painstaking development and tidy administration.

However, since 1938 the two main influences in my working life have been Jaguar Cars and Coventry Climax Engines. Nowadays, of course, they are both part of the British Leyland Motor Corporation, and I was happy to be able to control engine development at both companies in the 1960s, but when I first got to know them they were both thrusting, independent companies which eventually were to make great names for themselves. Of course I was involved in those famous 'fire-watching' discussions about the sort of engine Jaguar should make after the end of World War Two, and it was fascinating to be in the design and development team which produced the legendary XK engine; incidentally, how nice it was for me to return to Jaguar in the 1960s to find that the XK engine was going as strongly as ever. Latterly, of course, Harry Mundy and I devoted many hours to the V-12 engine of which we are now so proud, but both Harry and I are likely to be equally proud of our time together at Coventry Climax, when we seemed to design and build more engines of different design than anyone else in the time. But then Leonard Lee always liked to have new engines around him, and when it came to racing no man could have given me more encouragement. I was personally responsible for every racing engine that Coventry Climax designed in that wonderfully prolific period from 1952 to 1966 – though I doubt if this could have been achieved without first Harry Mundy's help, and later that of Peter Windsor-Smith and our excellent design and development team.

Can it really be 55 years ago that I joined Bentley as an apprentice? Was I really involved in things like the Barnato-Hassan Special, John Cobb's Napier-Railton Land Speed Record car, or William Lyons' SS Jaguars of the late 1930s? I suppose it has been a long time, but I can truthfully say that I am by no means tired of it all. I have often been asked if there was ever a period when I was unhappy in my work, or if I ever thought I had made a mistake by moving to a particular job. Although I was with some companies for only a short time, I have always been happy, and have gained friends, contacts and useful experience at all times. Indeed, I can say with all honesty that I have enjoyed every phase of my work since leaving school, and I look back over my career and am content with what I have done and what has been achieved. On the whole I would not have wished for anything very different, and the nicest comment of all is that I am still involved in motor sport. Why on earth should I want to stop!

Leamington Spa, March 1975. W.T.F.H.

1
Bentley Boy

My career in the motor industry always led me progressively from one interesting task to the next, yet my very first paid employment came about completely by chance. It was during 1920, after I had left the Hackney Technical Institute with a wealth of knowledge but without a diploma of any kind, that my father was looking around London for a suitable place for me to start as an apprentice. Jobs were very hard to find then, for after the war there was a recession in the engineering trade, and we had tried all sorts of places without success. Father knew the manager of the Sunbeam company's service department – which had just moved over from Lower Holloway to Hendon on the banks of the Welsh Harp – and one day we went over there to see him. He gave me the same answer as all the other companies – "We'll put you on the list of applications, but we don't know when you could start" – and we went away feeling very depressed. However, as we left, we looked down the road and saw a new factory being built in Oxgate Lane, Cricklewood, with a name on the wall. The name was 'Bentley', and it was to give me my first job.

I had stayed on at school longer than I needed to do by law, and I was 15 years old when I joined Bentleys. I was born on April 25th, 1905, and according to the official definition I am a cockney, because I was born within the sound of Bow Bells. My father was a buyer and the manager of a men's outfitting shop, part of a large store called Knowlman Brothers, in Upper Holloway, where we lived for quite a while, just below Highgate. In those days London ended not too far away from there, and from the Archway it was pretty well open country all the way to Finchley. Ours was eventually a large family, as most seemed to be in those days; I was the eldest of six children and had three brothers and two sisters, although the youngest of the girls died when she was quite young.

I don't know why I should have become interested in engineering, because father certainly wasn't, and the nearest we had to an engineer in the family was my grandfather, who was a chairmaker and cabinetmaker. My father's elder brother carried on in that business, while father always seemed able to make things – quite reasonable things – with the minimum of tools. I suppose my first interests were aroused through my cousin John, who was about three years older than me. He was a keen modelmaker, and through him I became interested in magazines such as *The Model Engineer* and *Amateur Mechanic.*

When I was five years old the family had moved into a flat in Finchley, and I went to a succession of schools including the East Finchley Grammar. Then came the first war, father joined the Royal Flying Corps, and the family finances were strained quite a lot. I had to leave the grammar school and go back to an elementary school for a while, but after a couple more years I was put into the Northern Polytechnic in Lower Holloway. By then it was becoming obvious that I wasn't at all interested in selling socks and shirts like father, and had become more interested in the engineering and other science classes, so at the end of the war father fixed me up at the Hackney Technical Institute, which was a full-time engineering day school.

I stayed at Hackney until September 1920, and because I wasn't very good at examinations then - I suppose one could call me a late developer - I never achieved any scholastic honours. In 1920 there was no easy way of becoming established in the engineering profession because of the trade conditions. If one looks at the motoring magazines of 1919 and 1920 it was obvious that there were dozens of little companies who had decided to make cars, but it didn't take long for many of them to go bankrupt. It wasn't that they all made bad cars, but there simply wasn't much money to spare at the time, and unemployment was high. So getting me started in a job wasn't easy, especially as I really had a liking for marine work, in which father had no contacts at all.

We didn't have a car in those days - no-one of our class could really afford to go motoring just yet - but I had come into contact with motor cars and trucks through a man called Charlie Dawkins, who lived near us in Finchley. During the war he had been in charge of the engine test shop of a company called Tylers, who had premises near Kings Cross Station, where they made engines for Karrier lorries intended for the Army. Of course that was many years before Karrier became part of the Rootes Group. These particular Karriers were known as 'subsidy lorries', which was a name given to vehicles built to a Government specification which made them suitable for military use, and for which the manufacturer received a subsidy. Dawkins sometimes took me down to his factory at weekends and he used to let me go into the test beds and watch the engines being run. He also let me have spare copies of various blueprints - very impressive they were too - which really began to develop my interest in this sort of thing. Dawkins also gave me my taste for marine engineering, because Tylers made engines for the Lifeboat Institution, and he used to tell me about the trials he had to make at sea to be sure that the product worked correctly. In the end he lost his life, by drowning, when on this work.

However, there was no way that I could get started as a marine engineering apprentice, nor with any of the railway companies. We tried the Great Northern and the Great Eastern but got the same response as everywhere else - there were no vacancies and the personnel people simply couldn't promise when the next one would materialise. Father then turned to the motor industry, and we tried AEC in Walthamstow (they're part of British Leyland now, at Southall), then Daimler, and then Sunbeam.

When we came away from the Sunbeam service depot in Hendon we were both in a pretty disconsolate frame of mind, but the sight of that interesting-looking new factory going up on the other side of the road provided us with at least a glimmer of hope.

We wrote a letter of application to Bentley, not knowing much about the company at all (in 1920 the Bentley car had still not gone into production). To my joy we were asked along for an interview, which was conducted by a man called Frank Clement, who was then in charge of all the experimental work. The interview was successful, and I was appointed as a 'shop boy' at the princely wage of ten shillings (50p) a week. It sounds like a ridiculous wage now, and even then it was precious little, but at least I now had a job, and of course it was a real stroke of luck for me because Bentley was destined for a wonderful future; I really began to learn the business from the ground up.

The number 14 on my clock card meant that I was the fourteenth employee to join the company at shop-floor level. The working day was from 8 am to 5.30 pm, with a lunch break of an hour, and of course Saturday morning was also compulsory. Because of the experimental work we were doing there was a lot of overtime to be done, and much later Frank Clement used to say that one of the things he could best remember about me was that I was so keen to do overtime work that I often exceeded by a large number the maximum number of hours allowed for young persons. After a factory inspector had come around to check up on these things I had to cut down on it; of course it all helped to give me more money, but it wasn't only for that – I was genuinely interested in the job.

When I arrived at Oxgate Lane there didn't seem to be much evidence of activity, in fact there were just two prototypes and several engines. Of course, Bentley Motors had only been founded in 1919. W.O. brought in the head designer from Humbers, F. T. Burgess, and along with Harry Varley they got down to the job of designing the first Bentley car in an office at Conduit Street, in the West End of London. There was never any money to spare, and in the early days W.O.'s brother, H. M. Bentley, concentrated on keeping things afloat by selling as many French DFP cars as he could from the showroom in Hanover Street, just round the corner. As soon as the engine design was finished – the famous 3-litre which went on to win countless races, including two at Le Mans – the first mechanic, Nobby Clarke, moved into the New Street Mews shop, over DFP's service workshops, to start building it up. It's worth mentioning that another of the early employees was Clive Gallop, who had worked for Peugeot before the war, and who had a lot to do with the design of the camshaft and cylinder layout of that first engine. W.O. certainly didn't design the whole car himself, nor even all the engine, but he was the driving force behind everything that happened there.

When the first engine was started up in the Mews, staff from the nearby convalescent home rushed round to complain about the noise, so it was fortunate that the Oxgate Lane site was soon purchased. Heaven knows where the money came from (though W.O. had received a sizeable award for his successful BR – Bentley Rotary – aircraft engines, but perhaps nearly all of it went into the land because there never seemed to be any to spare for equipping the shops!

Oxgate Lane at that time was at the very edge of a sparsely built-up area in north-west London. There were no factories beyond us, just open country all the way to Hendon, and there was only a country lane linking us with the wood-blocked and very slippery Edgware Road. It's all a bit different now, of course, with the built-up area stretching as far as Elstree, but most of our original buildings remain on what is now the site of the Smiths Industries instrument company.

We used the Edgware Road as our test route because apart from West Hendon, at the top of the hill beyond the Welsh Harp (now known less romantically as the Brent Reservoir) there was clear road nearly all the way to St Albans.

To get to work I normally used a bicycle, along the country lane from Finchley which is now the North Circular Road. If I hadn't used my bike I would have had to walk about two miles from home to Church End, Finchley, then catch a tram or bus to Golders Green, then another one to Cricklewood, and yet another up the road to the factory.

When I started there, the factory where the cars were to be built was still unfinished, and all the experimental work including engine testing was going on in a tall brick building which was the first to be completed. The money situation was so tight that when W.O. very rightly insisted on buying the latest Heenan and Froude water brake in order to test the engines, other directors accused him of being grossly extravagant! At first we had only two prototype cars – W.O.'s own and the first experimental car which originally had been an open four-seater, but was now fitted with a very sporty two-seater shell with the spare wheel strapped on the rear. One of the first important milestones was the acquisition of a lathe – this was very big news because when I started we had one electric brace, a portable hearth, a big paraffin blow-lamp and precious little else! All the machining was done outside by specialists.

Soon I was so busy, and getting more and more involved in Bentleys (it was that sort of firm – even though the money kept threatening to run out, somehow we always believed it would continue), that I had no time to go to nightschool as originally intended. However, after a couple of years father pressed me into going to the Northampton Institute in Islington. But even that didn't work for long because I became so involved in more interesting Bentley work that I dropped out. It wasn't until Bentley's chief draughtsman, Mr Dewhurst, who was much involved in the Institute of Mechanical Engineers, persuaded two other youngsters and myself to go to the Regent Street Polytechnic that I really became serious about higher qualifications. We stuck at this, and eventually I passed the Institute's Graduate Engineers' examination.

It was not easy, though, because I had to attend three evenings every week. That meant getting on to the tube train to Regent Street among the theatre-goers, and coming back among them, too. Washing facilities at Bentley were pretty primitive - we only had buckets of cold water and Hudson's soap powder to get off the day's filth before we set off - so I could never say that going to night school was fun.

I had hardly any leisure life at all, but the job was so exciting and full of interest that I never really noticed the tremendously long hours. I always looked forward to going to work, and when I was there I enjoyed it to the full. People used to ask me if I noticed a special atmosphere at Bentley, but as I had never worked anywhere else I couldn't possibly comment. Certainly all the people who worked there were hand-picked - there was plenty of choice open to management in those days because of the high unemployment - and we were all very keen to build up a new company around a man who was already well-known and respected. As soon as the cars started to win races the mystique built up rapidly, and just as Bentley became a very fashionable car to own, so Oxgate Lane became a very fashionable place in which to be. I know now that there was a very special atmosphere in the company.

The first production cars were delivered towards the end of 1921, and almost immediately after that W.O. decided that we would become involved in racing. He had raced DFPs before the war, and of course so had Burgess, so it didn't really need much persuasion. The first cars were to be built up in our shop, and I became involved almost straight away. Even though I was only 17 years old, I was taken away on the company's first racing trip - which was to the Tourist Trophy race in the Isle of Man in 1922. It didn't take me long to realise that there was more than racing to a trip like this. No sooner had we become installed in our hotel in Ramsey when the older mechanics took up with the most enterprising local girls, and went off for evenings with them. One night, Wally Saunders, Jack Besant and Les Pennal arrived back very late from a dance to find the hotel doors locked and barred - even the garage doors were locked. The only way they could get in, they thought, was by wakening Wally's father, Arthur, by throwing pebbles up to his window. This worked OK, as did the traditional method of making a rope from blankets, but Arthur laughed so much when he was hauling up his son that he let go of the 'rope' and down it came, complete with Wally. After that they decided to abandon the mountaineering, so scrambled over the garage wall and spent the rest of the night in W.O.'s saloon car!

After that I became progressively more involved in racing, and apart from the service work which I did during the winters, when the cars were not being raced or rebuilt for the next event, I remained with the racing team from 1922 to 1930. We had to teach ourselves about the detail preparations for racing in those days, as it was really breaking new ground after a four-year world war, and in any case we were all very new to the game at Bentley. However, we soon gained a reputation for putting cars together properly. This is something I have always remembered – the very fastest car or the most powerful engine will never be successful unless it is built and prepared properly, and is reliable. Preparation of those cars took a lot of time, especially as we had little in the way of machines and had to do much of the work by hand. For weeks before we had the new lathe, as the youngest lad in the shop I had to supply the power! I used to wind on an old mangle arrangement which worked various drills and cutters, and I always seemed to qualify for any odd jobs which were going.

I didn't go to Le Mans in 1923, the first year the 24-hours race was run, because the Bentley which started (and made fastest lap, though it only took fifth place overall) was privately entered by John Duff. Duff, who was born in China, became one of Bentley's London agents, and had already started racing at Brooklands, where he took the British 'Double-Twelve' record in a short-chassis 3-litre in September 1922. The same year he also distinguished himself by failing to pull up quickly enough after a 100 mph dash down the Finishing Straight in a big Benz, and went straight over the top of the Members' Banking at the end of it! But it didn't seem to cramp his style, even though he broke his ankles, because he abandoned the written-off Benz and raced his Bentley at the next meeting!

Duff had some work done on his Le Mans car in our experimental department, and he took along Frank Clement as co-driver for the race. Surprisingly, W.O. was against the whole idea of 24-hour racing at first, but he was persuaded to go along to watch, and after a few hours became quite excited by the event, and even more enthusiastic about the possible publicity he could get for his cars by winning. The following year Duff entered his car again, and this time there was full support from the works. Les Pennal was really the first permanent racing mechanic, and was given a corner of the 'finished car shop' where he laboured on the car for months before the race. This time there was no mistake, for Duff and Clement won the event, though their speed was a bit down on 1923. Clement was a remarkable man. He worked for the company from start to finish, and he also drove at Le Mans every year from 1923 to 1930, with considerable success.

My first trip to Le Mans was in 1925, the first year the company entered cars officially, and I suppose I really distinguished myself that year by breaking the rules without any officials finding out. Both cars ran out of petrol before the end of the first 20 laps, because no-one had thought about the extra drag which would be caused by having the hoods up (it was compulsory then, for the initial 20 laps, although that particular rule was soon to be dropped) and of course the only place they were allowed to refill was at the pits. I remember Duff went missing for quite a time, then suddenly appeared, running, having abandoned the car out at White House Corner. He wanted us to give him some petrol so that he could get the car started again, but as it was surrounded by people it wasn't going to be easy. Eventually I found a push-bike and offered it to him along with a bottle of petrol, then I found that he couldn't ride the bike because of his height and so I had to go along with him. Once we got to the car we fiddled about among a crowd of gesticulating Frenchmen, unscrewed the top of the Autovac fuel pump, and managed to pour enough petrol into it to get the car started and back to the pits. Then I rode back on my bicycle, and no-one ever found out what had been going on. Not that it helped, though, because later the car broke down when one of the carburettors broke off and started a fire.

That same year I became involved in a 24-hours record attempt at Montlhéry, where Duff and Woolf Barnato drove a very special 3-litre. This was the car which had taken the 'Double-Twelve' record at Brooklands at about 86 mph in 1922. At that time it had a really spartan four-seater body with just one steel bucket seat with no trimmings except a cushion. After the first 12 hours Duff had been so stiff that we practically had to lift him out of the car. His back had been rubbed nearly raw because the peak of that bare seat came just below his shoulder blades – and at Brooklands the car gave an extremely harsh ride because of the rough bankings. For the next 24-hours attempt with Barnato, Duff decided to use the Montlhéry track near Paris because it was known to be that important little bit quicker than Brooklands and because the surface was so much better. It was a shorter track, too, a pure oval, and this meant that the pits saw quite a lot more of the car and could keep a better eye on it. Les Pennal and myself were sent over to Paris to the Weymann factory to carry out the installation of their special streamlined body for the car. That attempt was successful, for Duff and Barnato took the record at 95 mph – which showed that the Montlhéry track was really quite a lot faster than Brooklands.

I never made it to Le Mans in 1926, not because I wasn't chosen to go, but because I had injured myself at Montlhéry! In fact, if it hadn't been for prompt attention by Wally Saunders, one of my Bentley colleagues, I might not have survived a very nasty crash. W.O. had set his sights on getting 100 mph from a 3-litre for 24 hours at Montlhéry. For this attempt we set about modifying the special 3-litre sprint car that Clement had built for Brooklands racing and other sprint courses. Dr. J.D. Benjafield had bought it from Clement, who had used it to equal the hill record at the last meeting ever held at Kop Hill. Benjy lent it to the company for the record attempt, and we rebuilt it as a single seater. It already had a short chassis – with some 12 inches cut out from the frame and plates over the joint – and on it we built a single-seater Gordon England body made of canvas, thin wood, glue and nothing else! It was shaped just like an aeroplane wing in profile, with the deepest section over the radiator and engine, and the body sloping away around the tightly built cockpit to a long, low, pointed tail. The body was so stiff in torsion that it did not match up very well with the flexible chassis. The springs were stiff, the axle movements were very restricted, and the constant flexing as the car ran on to and off the steep bankings caused the body and its mountings to break up. In the end we had to

W.T.F. Hassan, OBE—a recent portrait

This is where the Gordon England-bodied 3-litre Bentley record car finished at Montlhéry after my crash in 1926. I was thrown out when the car rolled towards the camera position

How we used to motor at first! Ethel and I in the overhead-cam Morris Minor, with fabric bodywork, that served me for several years

devise a form of three-point attachment of the body to the chassis, with a single central mounting point at the rear.

We made several abortive attempts on the record, interrupted by engine failures and various other problems, and it was a long time before we were ready to have a real go, by which time some of us were wondering if the car would succeed after all. It was June 1st, 1926 - not all that long before Le Mans - when the car set off on what was destined to be the final attempt. Clement, Barnato, Benjafield and George Duller were to drive, and Kensington Moir was the reserve driver; W.O. was there as well, but not expecting to have to drive. Duller started and did quite well, but soon the weather turned really nasty, with torrential rain and gales, and after all the drivers had taken their turn it was quite obvious that none of them was enjoying it. It was well after midnight that Duller was due to take over again - we had now been going for well over 12 hours - and when Barnato pulled in after his stint, really tired, Duller was the only driver in the pits. Clement and Benjafield had gone back to the nearby chateau to get some rest, and then W.O. went off with Barnato. That left just Duller in the car and Pennal, Wally Saunders and myself in the pits.

One gets into a very fixed routine at a record attempt. The car is expected to go past the pits regularly, and at Montlhéry its droning engine note could be heard all round the track from the pits. So when the noise suddenly stopped we wondered what on earth had happened, and naturally we feared the worst. So Pennal and Saunders set off in the dark and the rain in search of the Bentley, driving our old Morris tender car the wrong way round the track. They found it at the nearside edge of the circuit, where Duller had ended up after spinning it a couple of times on its smooth tyres. He had regained his bearings and restarted the engine when they arrived, so he got going again and reached me in the pits, where of course I was all alone. He told me that he had hurt his shoulder and neck on the cowling during the spins. He was dazed and shaken, and climbed out of the car saying that he was chucking it in. I never really hesitated. I was so enthusiastic in those days and so anxious to get the best out of everything for the company that I just jumped into the car, put on Duller's helmet and goggles and set off. I must have been mad, because not only had I not driven the car before, but I was quite unfamiliar with the track. But that didn't stop me.

It was still raining hard and the tyres were quite smooth, so I only got round the first banking, out of second into third gear, and then the car spun round, went straight through the barrier blocking off the exit road to the road circuit, and rolled over into a ditch. Afterwards the others told me that the car had rolled two or three times and finished up on its wheels badly damaged, but that I had not been thrown out. I was tremendously lucky that the car seemed to roll in such a way that my head didn't get crushed underneath it.

Wally Saunders and a French Shell representative called Paul Dutoit heard the crash and rushed round to look for me in the old Morris. They found me laid out in the car with my head against the back where the streamlining had been, and as far as they could see I had stopped breathing. Apparently the Frenchman said something like " 'E 'as cooked 'ees goose!" and thought I was a goner. Eventually they dragged me out of the car (which couldn't have been easy because I was all arms and legs, very lanky), got me into the back of the Morris, and set off towards the village of Linas to look for a doctor – the luxury of medical aid on the spot for record attempts and races didn't become commonplace for at least another 20 years.

On the way down the road, and one must remember that it was still dark, Wally noticed that my face was covered in mud, and he started poking around in my nose and mouth to clear them. Wally always said that eventually he dug out enough and suddenly I started breathing again, "just like unstopping a blocked sink". After that I was really not in any danger; my shoulder was badly knocked about, and I had a big cut across my lips, but I was concussed as well, and after the French doctor had knocked me out with chloroform I didn't really know anything else until I was in the Hertford British Hospital in Paris, recovering. So I think it is obvious why I didn't get to Le Mans that year; not that I missed an awful lot, as two of the cars retired with mechanical breakdowns, and Sammy Davis put the third one into the sand at Mulsanne less than half-an-hour from the finish, when he might have been second.

The following year it was quite clear that W.O. meant business at Le Mans. The effort started sooner, with new cars which received a lot of attention and special preparation, and we had a strong team – one of the new $4\frac{1}{2}$-litre cars (actually based on a 3-litre chassis) and a couple of 3-litres. One reason for the big effort was that we were still smarting over our failure to win the previous two years, but another must have been because of the new money which Barnato and his associates had put into the firm. As far as I can be sure (for I was never involved in the financial side of things) Bentley Motors got through three lots of money before they finally collapsed in 1931. The original Bentley capital was used up almost as soon as the first cars were built, and in order to keep going it was necessary to encourage motoring enthusiasts, business speculators and even motor traders to invest in the company. Woolf Barnato had been driving Bentleys for some time before he was asked to put money into the firm. This happened just at a time when the development and production of the big new $6\frac{1}{2}$-litre car was making it difficult even to pay the wages bill every week. However, in 1926 (which was the year of the General Strike, though that didn't affect Barnato's fortune, which had come from his father's business operations in South Africa) Barnato became Chairman and one of his associates, the Marquis de Casa Maury, became Joint Managing Director alongside W.O. With appropriate financial backing Barnato set out to operate a proper racing programme.

Almost immediately the racing department was moved out of the main works at Oxgate Lane to the Service Department at Kingsbury. We took over a workshop from Vanden Plas, who just happened to be next door, and who made all our special racing bodies. In this way I was divorced from the main activities in the factory for the whole of the racing season, which included not only Le Mans, but also the TT at Belfast, Phoenix Park, and long-distance events at Brooklands like the 'Double-Twelve' or the 'Six-Hour' races. In the winter, of course, it was a different story.

Somewhere along the way I had also managed to meet the girl who became my wife, but there really was a very long gap between our first meeting and our marriage in 1933. Ethel insists that the first meeting wasn't very romantic. She was then Ethel Murray, and she lived in North Wembley. Her father was a great motoring enthusiast and it was not long after I joined Bentley that Arthur Saunders (Wally's father, and one of the company's very first employees, and incidentally a driving mechanic in motor racing for Humber some years before I started that game) took a car up to Wembley to show to Mr Murray. I remember that it was a very cold day, and Arthur went into the house leaving me outside in this open car. While Arthur and Mr Murray were warming themselves up with whisky, Ethel's mother asked Arthur if he was alone? "No, no," said Arthur, "I've left the boy outside." As shop boy I was just about as unimportant as it was possible to be in a company, and Arthur would not have thought it appropriate to take me inside such a nice big house. Ethel's mother then said "Bring him in. Bring him in to have a drink!" Arthur then said, "He mustn't have a drink, Good Lord, he's only about 16. No, he's all right out there, he's just the lad from the works, the apprentice!" He might just as well have said "Let him freeze" I suppose. But Mrs Murray insisted that Ethel, who was a shy young thing the same age as me, should bring me out a hot cup of cocoa to keep me going. So this pretty girl came out and passed me the cup of cocoa; we looked at each other, but apparently had very little to say to each other. As I have already mentioned, I was so bound up in my job that I didn't find much time for a social life, and I suppose I didn't really know how to set about talking to a strange girl. I don't think we saw each other again for several years, but then Wally Saunders married one of Ethel's sisters, and I used to go round to the house at Wembley with them; it was only then that we started seeing anything of each other regularly.

So much has been written, painted and talked about the famous Bentley racing cars and the 'Bentley Boys' that there is no point in my repeating it all. However, I do remember that once we started building cars really seriously for Le Mans we never lost the race again; we won it four years in succession, from 1927 to 1930. The first time was when Sammy Davis and 'Old Number 7' won after getting involved in the White House crash which eliminated the other team cars, and the last time was in 1930 when Barnato's Speed Six won easily, having beaten off that 7-litre supercharged Mercedes that Caracciola drove so hard and well. I should also say that W.O. never allowed any of his cars to be driven faster than was necessary to win, so that the opposition would never really know what sort of performance they would have to beat next year! Only when Birkin went out to bait Caracciola at the start of the 1930 race did we ever see a Bentley really flat-out – and it didn't do him any good because he burst a tyre when he pulled on to the verge to pass! By the end we were really winning as we pleased, which showed up in the Le Mans entry; by 1930 everyone expected us to win anyway, and there were only 18 starters of which six were Bentleys!

At Le Mans we usually stayed at the Hotel Moderne, which had a good-sized yard and three lock-up garages. The bedrooms were strung out along two corridors with long strips of carpet on the floor. We mechanics were always out for a bit of fun, and on one occasion we decided to up-end Dudley Froy by pulling the carpet from under his feet when he came out of his room; naturally we got it wrong – when the door opened, the person we up-ended was Baron d'Erlanger, who was with him that year! He had quite a reputation for being poker-faced, and on that occasion certainly his face didn't slip. There was another equally notable time at Le Mans when he tried to start a Lagonda after the Le Mans running start. The engine wouldn't fire, not at any price, but the good Baron kept churning away at the switch, completely impassive, until the battery flattened itself.

Was life really so much more fun in those days, or is it just that I remember the most hilarious occasions? Surely no-one could repeat the incident of the goat any more. That was the time when drivers and mechanics all went off for an evening's entertainment to a rather notorious café. The madame paraded all her girls – yes, those sort of girls – for inspection, whereupon one of the most senior and respected members of our team was heard to say that he would prefer the company of a goat. It all went very quiet for a time, and we were sure the girls were offended, but after about half an hour madame reappeared – with a goat! The offer was not taken up.

We were very green in pit procedures at first, and Sammy Davis, who was already vastly experienced, took charge of training. In those days we used enormous petrol funnels to aid refuelling – so big that three churns could be thrown in at once. One day Sammy and his mechanic, Head, were showing us how to do this, very slickly and very efficiently, and once the tank was filled they drove off with a flourish. The only snag to a very impressive performance was that the funnel was still in the tank . . .

All the Bentley engines, with the exception of that rather strange 4-litre six-cylinder design we produced right at the end, were laid out in the same way, and by almost any standards they were good pieces of engineering. W.O. made no bones about studying all the best engines before he started on his own, and in particular he must have looked at the best racing cylinder-heads before he designed the first Bentley. I believe the Bentley engines were the only production units built in Britain for many years with four valves per cylinder, and in the light of present-day Grand Prix design it is extremely interesting to look at the layout again. There were two inlet valves on one side, two exhaust valves on the other, with two sparking plugs, one on each side of the cylinder-head *below* the valves. By comparison with the Duckworth-designed Cosworth

Ford DFV and the latest V-12 BRMs, the Ferraris and the Coventry Climax engines with which I was concerned, the only real change in the chamber is that the later engines had central sparking plugs! That wasn't possible on any of the Bentleys because they used a single overhead camshaft operating the valves via rockers, and there simply wasn't room for the plug to be in the proper place. In the 1920s, of course, we were saddled with the RAC's 'Treasury Rating' horsepower tax, which was calculated by reference to the cylinder bore only (not the engine's capacity), and meant that even the most powerful cars tended to have engines with small bores and long strokes. It cramped our engine designers' thinking for many years, and even after the second war we still used engines (like the XK Jaguar unit which I knew so well) which were well 'under-square'.

All the Bentley engines except the 4-litre were closely related – there was only a single basic design, though some Bentley enthusiasts might like to tell you otherwise – the biggest differences being in the way the drive to the overhead camshaft was arranged, and of course the number of cylinders. The first 3-litres used a vertical-shaft-and-bevel drive from the front of the crankshaft, but the six-cylinder engines which were developed from it used that rather ingenious system of coupling rods driving miniature crankshaft throws at the rear of the engine. It was a very interesting and complicated, but silent, method of doing things, but it increased the length of an already large engine by something like six inches. I could not see any cost accountant accepting that sort of thing in the 1970s! Strangely enough, when the $4\frac{1}{2}$-litre engine was designed, which was really an amalgam of 3-litre and $6\frac{1}{2}$-litre details, it reverted to the original vertical-shaft drive.

Bentley enthusiasts still write off the final 4-litre engine as a failure, but in engineering terms this simply wasn't true. It was the result of a lot of work by a lot of people; we ran several single-cylinder test engines at Cricklewood, and there were at least three experimental cylinder-heads – one from Harry Ricardo, one from Harry Weslake (who had already done great things to the $6\frac{1}{2}$-litre's breathing) and one from Whatmough. When they had all done their best, W.O.'s design team set to and incorporated the best of their combined efforts into a six-cylinder engine. This had a single, large, overhead inlet valve and a single side exhaust valve in a combustion chamber inspired by Ricardo's ideas. The plug was placed over the exhaust valve, ideally placed to promote good combustion, whereas on earlier W.O. designs the plugs were side-mounted in the cylinder-head, either side and underneath the valves and ports. When the first car was built we were all terribly disappointed with it, but that was really because the poor little 4-litre engine was made to pull an 8-litre chassis that had only been shortened by a foot, to give a wheelbase of 11ft 2in. As an engine it really was well designed, and it produced quite a lot more power than the $4\frac{1}{2}$-litre with half-a-litre less, *and* it was reasonably quiet. I remember that we proved this one day by loading up a $4\frac{1}{2}$-litre car to the weight of a 4-litre and testing both cars up Brockley Hill, north of Stanmore; of course, the 4-litre won. Only 50 4-litre cars were ever built, and most reckoned that all the money spent on the engine just hastened the collapse of the company, the assets of which were bought by Rolls-Royce in 1931, but later many very similar features were to be applied to their successful 'B' range of engines which appeared after World War Two.

Being a racing mechanic in the 1920s involved quite a lot of personal bravery, for not only did we build the cars and attend the races, but we rode in the cars during the races as well! Driving mechanics were compulsory, even in Grand Prix cars, for many years. Their main duty was to keep a sharp look-out behind, then warn their driver if someone else wanted to overtake. In those days, drivers weren't so ready to 'put their elbows out' to make it difficult for anyone else to pass – they gave way gracefully. We also had to perform any repairs and wheel-changing necessary away from the pits and – most important – we had to pump away to keep up the air pressure in the petrol tanks. None of us rode exclusively with any one driver – we just jumped in and out at pit stops whenever it was convenient. I always enjoyed riding as a mechanic because it was really the summit of achievement in our profession at that time. However, I must say that some drivers were a lot safer than others! It was during this period that I struck up a very happy relationship with Woolf Barnato. He might have been chairman of the company, but once in a racing Bentley he was no different from the next team man; he was extremely good at the wheel, and always obeyed team orders.

I rode in the 'Six-Hour' and the 'Double-Twelve' races, but I never rode at Phoenix Park or in the TT; by then I was chief mechanic, and I had given up my riding jaunts for an organising job in the pits. In really long events like the 'Double-Twelve' we swopped around, and I certainly rode in every Bentley car in one race. There were sometimes nine or ten Bentley runners, including people who had bought last year's cars, and we looked after them all.

I have two particularly fond memories as a driving mechanic. In the first of the 'Six-Hour' races we ran 3-litre cars with experimental light-alloy rocker gear, and they all started jamming. One car – Birkin's old one – had the original steel rockers, and I finished up in that because it was finally the only one running. Then it started to give trouble in the gearbox, and Frank Clement and I climbed into this car to nurse it round, but things got worse and worse. In the end I had to lift the lid off the gearbox and slog a selector into third gear with a big copper mallet. We finished the race in third gear!

My other outstanding memory is of riding with Birkin around the Nurburgring in a $4\frac{1}{2}$-litre when he was valiantly trying to beat a fleet of much bigger and faster Mercedes cars. It was a very hot day, with a very slippery road surface because the tar had melted in the heat, yet Birkin drove extremely quickly and well. I thoroughly enjoyed that day, even though he couldn't beat the big Mercs.

In the middle of the 1920s we had another life – a completely different job in the winter months. When the racing shop closed down after the last race of the

Le Mans, 1928. Bernard Rubin fills up the winning car as a French official seals the oil filler and Woolf Barnato shouts instructions. Next to Barnato, in order, are Nobby Clarke, myself, Stan Ivermee, Marquis de Casa Maury, W.O. and Frank Clement

Jack Dunfee and myself rolling into the Brooklands pits at the end of the 1929 Six-Hours race. This was the famous 'Old Number One' Speed Six Bentley, which had just won at 75.88 mph. Barnato is applauding and Stan Ivermee is standing alongside him

season, usually at the end of September, some of us used to be issued with Royal Enfield V-twin motorcycles, with sidecars, and be sent out on the road as travelling mechanics. Bentley owners were very keen to have their cars serviced and maintained by factory mechanics, particularly if they could boast to their friends that they had had a mechanic from the racing team to do the job! Our motorcycle combinations were similar to the sort of thing used by RAC and AA scouts, and were painted blue with 'Bentley Motors' on the side. We loaded up with tools, spares, gaskets, valves, pistons and that sort of thing, then set off with a list of addresses where owners were waiting for the 'Bentley man' to call. This procedure was then fairly new in the motor industry. During the summer Bentley had one or two mechanics on the road, but most of the work could be carried out in the winter. Rolls-Royce had travelling mechanics, too, but in a much more discreet sort of way; the Bentley was such a 'hobby' type of car that the owners never minded anyone knowing we were coming!

Each of us was allocated a particular area, and mine was the northern counties - Yorkshire, Lancashire and Scotland, mainly - so I had to cover a lot of ground. We would be given the first few appointments for work to be done, but telegrams or letters were sent out from Cricklewood, redirecting us to the next job. I used to set off at the end of September, and would not be back until Christmas time. I got to know every telegraph pole on the old Great North Road, and it always seemed to be raining or snowing! We had to be prepared to do anything at all to the cars, from routine maintenance to complete engine rebuilds. Decarbonising was a very regular job then, of course - nowadays the fuels don't give the same amount of carbon build-up but in the 1920s decarbonisation was carried out on just about every job. All the Bentleys except the 4-litre had a fixed cylinder-head, which meant that the whole block had to come off; it was nothing to see me standing astride an engine heaving away to pull the cylinder block off the pistons. Generally speaking there were no panic jobs; they had already been arranged and unless the owners had very good garage facilities we often used to work in the local Bentley dealer's premises. That automatically meant that other owners used to hear that the 'Bentley mechanic' was in town, and there was never any shortage of work to be done after that.

At the time I suppose I was earning about £3 a week - I was still unmarried - and when we were out on these long trips we were allowed 14 shillings (70p) a day for expenses. We could find ourselves very good commercial hotels, with a jolly good evening meal and a big breakfast, we could go out to theatres or the music hall, *and* have a drink, all out of our 14 shillings and without having to break into our wages. I once got into big trouble when I eventually arrived back at the factory because I had been away for more than three months and in all that time I hadn't drawn any of my wages. The company were not annoyed with me for not drawing my pay, but because I might be spoiling the system for the others, and someone might come to think that the expenses were too high! Certainly it was a nomadic existence, but I enjoyed it because in a way we lived on the fat of the land. Occasionally there were trips abroad, too – I went to Paris several times and I had one flight out to Germany in a three-engined German plane, then drove down to Munich with the Bentley owner.

About half the Bentley owners used to have chauffeurs, but to many people a Bentley was a sporting car which they preferred to drive themselves. There were titled owners like Sir William Younger, a brewing magnate in Yorkshire, and Sir Michael Nairn, who were among my 'regulars', but many were sporting types and quite a few raced. The nice thing for us was that they were all well off, and most of them made sure we had a good tip at the end of the job! I could never forget servicing a 4½-litre for Sir Basil Zaharoff's daughter in Paris, because she always gave me a crisp new white £5 note! Of course, a fiver was well worth having, bearing in mind the very first car I bought cost me only £4.50. I once did a big job for a man who owned mills in Bolton, and his tip to me was a roll of casement cloth. It was supposed to be damaged, but we never found the flaws, and when my mother died not too many years ago there was still some cloth unused, though she had made good use of much of it in the meantime.

Some people had so much money, and their chauffeurs so much spare time, that their Bentleys received quite remarkable attention. Sir Michael Nairn had three Bentleys and two chauffeurs; every morning the cars' floor-boards were lifted and the inside of the undershield and the gearbox case were cleaned with metal polish! On the other hand, one of my customers was a farmer in Coupar Angus, who didn't seem to use his 3-litre a lot. He used to come up to London for the Motor Show, and go up to the north of Scotland for his holiday in August, but apart from that he really only used it to drive into Coupar once a week on market day; with him it was only ever a case of cleaning plugs and setting-up the magneto. I really had to be a jack of all trades, and I had to tackle almost anything from tickling up the carburettors to rebuilding after a smash; it was wonderful experience.

When I was out on the road I had the company's motorcycle to ride, but during the racing season I was faced with the journey from home to the factory. Although I had started out with great pride using my Hassan-powered bicycle, as soon as I could decently afford one I bought a car. My first, a GN cyclecar, wasn't much to look at but at £4.50 it certainly wasn't expensive when I bought it from Rowland Smiths. In the enthusiasm of my youth I ripped off the old boat-type bodywork and substituted a very racy two-seater with the petrol tank forming the back and the spare wheel clipped on behind that. It was chain-driven, of course, and when first gear eventually packed up I ran it for some time on second and top! There was a thriving 'goodies' business even then, because I remember buying overhead-valve conversions for the little engine. I part-exchanged it for my next car, which was a little Gregoire, a French car with an engine having most of the valve gear exposed to the breeze, and I ran this for quite a time until the axle shaft broke. I never bothered to repair that, but used the engine in a boat I had purchased, and then found a Fiat 509. This was a delightful little car with an overhead-camshaft engine all properly driven by chain from the flywheel area, and I had a great deal of fun in it. Motoring in those days *was* fun, because the roads were fairly empty, even if their surfaces were sometimes not up to much, and the sort of car I could afford was small and usually well-worn! Eventually I rose to a Morris Minor, with overhead-camshaft engine and fabric body, which gave yeoman service for several years.

However, just as I was looking forward to the end of my tenth year with Bentley as chief mechanic to the racing team all of us at Kingsbury received a considerable shock. Quite unexpectedly, when the company was at the height of its successes with the Speed Six competition cars, it ran into severe financial troubles, due mainly to the slump which had been triggered off by the famous 'Wall Street crash'. Of course this meant immediate curtains for the racing programme, and it soon looked very serious for the whole company. This was tragic, especially as the cars had been selling well, and were clearly superior to almost everything else on the race tracks. Sammy Davis summed up our racing position in *The Autocar* just after the announcement:

"The Speed Six team had reached a point where, if they did win a race, they got no credit because everybody said they ought to have won, and if they didn't win every ill-wisher in the place made as much capital of it as possible, which is a pretty sure sign that it's time to stop and let someone else have a chance for a race or two."

By 1930 I was completely immersed in the racing scene, and the withdrawal meant that I would have to consider a change of job. But at 25 years old I was not despondent, and I looked forward to the next interesting challenge which my life would bring.

Bentley Cars 1919-1931

The Bentley company's fortunes have been splendidly documented, not least by W.O. himself in his autobiography. However, it is interesting to recall that Bentley learned the basics of his craft in the railway industry, and later both raced and sold French DFP cars up to the outbreak of World War One. During the war he designed two successful rotary air-cooled aircraft engines – the BR1 and BR2 – subsequently using the Government's gratuity to help finance his own company.

He decided to make expensive well-engineered sports and touring cars when the war was over, and had the first prototype running in 1919. The 3-litre engine, his first car engine design, settled the basic layout of all Bentley units for the next decade. Common features were the use of four valves per cylinder, operated by a single overhead camshaft, a rigid crankcase, and a cylinder block and head cast in one piece. The last feature ensured that there would be no cylinder-head gasket failures by dispensing with one altogether, and it made the unit very strong and suitable for tuning for racing at a later date.

The basic layout must have been influenced to a great extent by the Grand Prix racing successes of Peugeot and Mercedes in the 1912-1914 period. The Bentley engines were the only 'four-valve' units sold to the public between the two world wars. The four-cylinder 3-litre engine suffered from a very long stroke (149 mm) and a relatively slim bore (80 mm), mainly because of the exigencies of the hated 'Treasury Rating', which imposed an annual Road Fund tax of £1 per horsepower, where the horsepower was calculated with reference to the number and bore of the cylinders but not to their piston stroke.

Three-litre Bentleys were sold in three different wheelbase lengths, and although their heyday was from 1922 to 1926 a few were sold as late as 1929. But several years earlier, work had started on a new car which was to have a six-cylinder engine, and in 1924 the first prototype was built with an engine that owed a lot to the 3-litre except that the stroke was reduced to 140 mm and the overhead-cam drive was by coupling rods at the rear of the block instead of by vertical shaft drive at the front. The engine's capacity was 4,224 cc, but tests showed that there was not enough power or torque. To overcome this deficiency the cylinder bore was increased to 100 mm, and the capacity to 6,597 cc, and after more development the 6½-litre Bentley was unveiled at the 1925 Motor Show.

Deliveries of the new model began in 1926, but by this time development of an earlier and larger four-cylinder engine, the 4½-litre, on which the new six-cylinder engine had been modelled, was under way. To rationalise production as much as possible the 4½-litre and the 6½-litre shared the same bore and stroke, the four-cylinder version also incorporating an amalgam of 3-litre and 6½-litre components, including the original vertical shaft-drive to the camshaft. The first 4½-litre car was raced at Le Mans in 1927 without success (it was involved in the notorious White House crash) but it won the 24-hours Grand Prix de Paris at Montlhéry later in the year, and a 4½-litre won at Le Mans in 1928, the year the model went into series production. Speed Sixes won twice at Le Mans, in 1929 and 1930, the two years during which this competition version of the 6½-litre remained in production.

The special 'Blower' Bentley was made in 1930 and 1931, being a version of the 4½-litre car with an Amherst Villiers-designed Roots-type supercharger mounted between the front dumb irons and protruding underneath the radiator. It was a car of which W.O. never approved, and all 50 of them were made in Birkin's own works at Welwyn. W.O. might even have been right; in spite of the extensive work that Villiers did on the engine, the car was never a success and never won a race in its prime. Nevertheless, this is the Bentley most often drawn, sketched, painted and modelled, even though the Speed Six was arguably the most successful.

The last great Bentley was the 8-litre, of which only 100 were made in 1931. The engine was an enlarged version of the 6½-litre (bore of 110 mm instead of 100 mm), but there was a very strong new chassis frame with optional wheelbases. The car was intended to match the best that Rolls-Royce (with the Phantom II) could make, was equally expensive and quite a lot faster. W.O. only approved of formal closed coachwork on this car, which could reach more than 100 mph so equipped, making it one of the very few production cars capable of exceeding three figures in 1931. Quite a few of these cars have been rebodied as 'replica' Speed Sixes, and several Speed Sixes have subsequently acquired 8-litre engines, although none was ever built that way originally.

The final Bentley was the least successful, even if its engine was not a complete flop. The 4-litre six-cylinder engine was the only Bentley design with a side camshaft and two valves per cylinder, which has led some people to suggest that W.O. never had a hand in it. It is true that W.O. never liked the design, and that several other consultants contributed time and expertise to it, but the final design was from W.O.'s drawing office, and incidentally the specific output was rather superior to that achieved by the first and most famous engine, the 3-litre. Unfortunately, the 4-litre suffered the insuperable problem of having to live in a barely-shortened 8-litre chassis, and usually had heavy bodywork to suit, so it is no surprise that it had a reputation for being gutless.

The Bentley company went into liquidation in 1931, having gone through three different amounts of finance, and after a rather sordid battle was taken over by Rolls-Royce, who had bought control through the medium of the British Central Equitable Trust, their nominee company. Bentley cars of the 'W.O.' type were never built again, and although W.O. joined Rolls-Royce as an engineering consultant, he had very little to do with the design of the $3\frac{1}{2}$-litre Bentley (often referred to as the Rolls-Bentley) that took its bow in 1933. Later, W.O. took the post of Technical Director at Lagonda, producing two more famous engines, the V-12 and the 2.6-litre twin-cam engine that later powered the DB2, DB2/4 and DB3 Aston Martins of the 1950s.

2
Barnato and Brooklands

After the Bentley company quit racing in 1930, I went back to experimental work and kept in touch with the service people, but soon it became clear that the firm was in desperate trouble. The money which had been put in to keep us going in 1926 had not been sufficient, and now the economic conditions that had followed the Wall Street slump were making it almost impossible to continue. Of course, W.O. didn't help by wanting to make only the best and most expensive car in small numbers – exactly the sort of car to suffer when business turned down and the rich men had to pull in their horns – but I doubt if the company had either enough capital or a sufficiently determined management to keep going anyway. Barnato decided that enough was enough – for most of the financial problems were his – and so the company was forced to call in the receiver.

However, no doubt because he had got to know me quite well, as I often rode with him as a riding mechanic at Brooklands and elsewhere, Captain Barnato invited me down to his big country house, Arden Run, near Lingfield, in Sussex. There he told me that although he was no longer Chairman of Bentley, he was still on the Board of Directors of Rolls-Royce which had bought the company, and he knew that not only would there be no more Bentley cars as we knew them, but there would be no more racing, either. He was still a rich man, and still very interested in motor racing, and he offered me a job privately, to look after his own racing cars and perhaps build one or two new ones. This sounded very interesting, particularly as he wanted me to operate from his workshops on the estate.

It was a really beautiful place, although it was about 35 miles from our home in North London. But Barnato knew I was planning on getting married to Ethel, and he said there was a cottage in the grounds of Arden Run which would go with the job. I don't really remember how much he proposed to pay me, but I think it was £5 a week, on which we could be comfortable, especially with the cottage he was providing. He was also prepared to pay my travelling expenses until I moved in – say from Finchley to Brooklands, or Lingfield to Brooklands, or any other journeys I had to do for him.

Celebrating the victory of 'Old Number One' in the 1929 Six-Hours race at Brooklands. Jack Dunfee and Woolf Barnato are in the car while I am walking, head down, past the cockpit

Wedding day—and didn't we look smart! Spats were *de rigour* then

I was still only 26 years old, so I don't suppose it worried me that I might be moving into a job which really offered little security. After all, it only needed Barnato's whim to change for him to pack up motor racing altogether, and I would be out of a job again. But then, there didn't seem to be any interesting future for me at Bentley. There had been that rather sordid battle for the company's name – we all thought Napier would buy Bentley and W.O. was already designing a new car to replace the 8-litre, but in the end a nominee company, an investment trust, bought the assets. It was only afterwards that we realised that it was Rolls-Royce which had bought control in this way, and it wasn't long before we also realised that they were not interested in making Bentley cars at all. Once the Cricklewood assembly shop stopped making cars it never started up again. Most of the good machine tools were shipped off to Rolls-Royce at Derby and the rest were sold.

Apart from the tragedy of the company's physical assets being broken up, it was sad the way the staff was also split up. W.O., of course, was taken off to work for Rolls-Royce – it was all part of the contract I believe – but a lot of people like me just left, although Rolls-Royce did absorb a considerable number of the old Bentley staff into their service department and their London workshops. I suppose we were all sickened at what was happening to our old firm – and, believe me, we all thought of it as 'ours' – and I was very glad to be moving straight into another interesting job, particularly one connected with motor racing and my old Chairman, Woolf Barnato.

I had found W.O. himself to be quite an extraordinary man. He always kept himself aloof from everyone – management, drivers and men – and possibly that was why he was so much respected on the shop floor. All W.O.'s staff were hand-picked, from designers down to the shop boy. He had a gift for assembling the right sort of people, and I think the results in just ten years confirmed this.

But now I was to work for Woolf Barnato, and at first I had to travel down from Finchley to Lingfield every morning, and back again in the evening – a round trip of about 70 miles, which must have taken three hours out of every day, even though there were no speed limits in 1931. The trouble was it was built-up area all the way from home to Caterham, and in any case the sort of cars I could afford would only cruise at about 40 mph even if the road was clear. Fortunately, by then I had qualified at the Institute of Automobile Engineers, so there was no more night school and consequently I had quite a lot of time to spare.

At first there was no suggestion of my building a special track car for Barnato; I simply maintained his road cars whenever they were based at Arden Run, and concentrated on preparing and developing his racing Bentley. The first car was the old Speed Six which had brought Barnato so much success in the works team, and it had all the successes he had gained with it engraved on the radiator. It had become his own property, and in October 1931 we took it to Brooklands for the BRDC's 500-Miles race – a real flat-out blind that was –

where Jack Dunfee and Cyril Paul were to drive it. Like all the other Brooklands races, this was run on a handicap basis – very difficult indeed for a casual spectator to understand because it wasn't often that a big fast car won just because it was fast. This time, though, Dunfee and Paul kept plugging away and eventually won at an average of 118.39 mph. At that time it was the fastest-ever winning speed for a long-distance race anywhere in the world – faster even than Indianapolis!

To many people it looked as if Barnato had a lot of cars, as the Bentley often turned up looking entirely different from the time before. Actually it all depended on the regulations, and what we could get away with. We had the old standard four-seater sports body for some races, a two-seater 'Brooklands' body at times, and a single-seater at others. It was the single-seater which we used to win the 1931 500-Miles race. Changing the bodies was a piece of cake, really, because we always kept the driving seat and all the controls in the same place, and of course there was plenty of room to hang up the spare bodies at Arden Run when we weren't using them.

Eventually we decided to retire the Speed Six, 'Old Number 1', and build a special track car, although the decision was rather forced on us when Jack Dunfee took it out in the Empire Trophy race early in 1932 and broke its crankshaft. Now I won't say that that sort of breakage was unheard of, but for it to happen to a Speed Six meant that the car had endured rather a lot of flat-out motoring. You could forgive it almost everything, though, because it had won a lot of races for Barnato, and a lot of prestige all round.

The new car was to be a purpose-built racer, and it was here that I put my ideas to work on an entire car design for the first time. We had encountered chassis-frame troubles on 'Old Number 1', so we decided to start with the strongest possible chassis. Although the 4-litre Bentley never had much of a reputation as a production car, its very strong frame, being a shortened version of the 8-litre, seemed to me to be ideal for the job. At first we put the rebuilt $6\frac{1}{2}$-litre engine out of 'Old Number 1' into the new car, but it wasn't fast enough, and somehow Barnato was able to get an 8-litre from Rolls-Royce. The 8-litre engines were *very* rare by then, so it needed considerable influence to get one out of Bentley's new owners.

There was no problem in installing the 8-litre in place of the $6\frac{1}{2}$-litre because physically they were exactly the same size. Although they were very large engines, it was not too difficult to get them out and back again with a small block-and-tackle; the crankcase and many other castings were in electron – very light and very expensive – and this helped considerably to keep the weight down. Barnato had an idea that he would like to use the car as a very fast touring model as well as at Brooklands, so we built it with a rather luxurious two-seater open body and a streamlined tail. We had no idea what sort of power a race-tuned 8-litre put out in those days, but as the production engine produced well over 200 bhp it could have been around 250 bhp. Even with that frontal area (we kept to a standard radiator because the engine was so high anyway) it was a very fast car, and we were hoping to do well with it.

It was entered for the 500-Miles race at Brooklands at the end of September, and was much the biggest-engined car in the race because people were now turning to small-engined cars to try to take advantage of the handicapping system. Jack and Clive Dunfee were to drive it. Jack was completely used to this sort of performance, but Clive hadn't really driven anything quite as fast, nor had he raced at Brooklands for more than a year due to his recent marriage, and he had done little or no practice. Racing in the 500-Miles was always a problem because of the enormous speed difference between our Bentleys and the little 750cc side-valve Austins and 850cc MGs. Passing was always hazardous because even the smallest cars used to run quite a long way up the banking which didn't leave much space for cars like ours, the big Delages and the Panhards. You get an idea of the speed differential from the fact that Jack Dunfee began *his* 500-Miles race one hour and thirty-three minutes after the slowest cars had started!

He would have to average well over 120 mph to win, and he started out at something over 127 mph. Jack handed over to Clive after a fuelling stop and wheel-change. Unfortunately, it wasn't long before the car suddenly went missing, and from the pits we were pretty sure we had seen it go over the top of the Members' Banking. By the time we found the wreckage there was nothing we could do for poor Clive Dunfee, who had been flung out and killed instantly; what was left of the new track car was down through the trees on the entrance road below. Barnato was horrified, of course, as were we all, but at least we were able to establish, later, that it wasn't car failure that had caused the crash; I'm pretty sure that what happened was the result of Dunfee's inexperience with the car. After the Vickers sheds one had to get the car really high on the Members' Banking, because soon after it started there was a hollow which caused cars to run down, then when they reached the end of the hollow they tended to run back up. I think the big car twitched down as usual, then back up again, at which point the driver didn't, or couldn't, stop it drifting up further. We knew from the marks that he had fought it along the lip for some distance, but at that speed there wasn't any hope and the car went over the edge. There was no retaining wall at Brooklands, ever, but in those days no-one made a fuss about it. Members of today's Grand Prix Drivers Association would have refused to drive at Brooklands under any circumstances, I'm sure, but if they'd seen the top of the bankings they would have had a fit and would certainly have refused to drive there before a retaining wall had been built.

Barnato kept what was left of the car for some time, and did nothing with it, then eventually he decided that it should be rebuilt as there was not a lot of serious damage. I remember that it needed a new front axle, wheels and so on, but eventually we rebuilt it and had a road-going coupé body rather like that of the existing SS put on it, with a longish luggage boot strapped on the back. It was a lovely car, and it went like stink, but we simply couldn't stop the exhaust fumes getting into the cabin, and eventually Barnato became tired of this and sold it. He never had any trouble in selling his cars, of course; all he had to do was let Jack Barclay's showroom in Berkeley Square take the car, they would put the 'ex-Barnato' label on it, and it would be gone in no time at all!

It was in 1933 that my life changed in several ways. For one thing I was no longer faced with the daily trek to Arden Run, secondly I started building the first 'Hassan Special' from the ground upwards, and third but by no means least, I got married. Ethel and I had been seeing each other for some considerable time, and once I had settled down in the Barnato racing operation I was feeling quite established, so we decided to marry. Some of my friends had been pulling my leg, and suggesting that it was taking an awfully long time to happen, and wasn't I being rather reluctant, but there never seemed to be a hurry. Anyway, eventually we made it, on Sunday, April 16th, 1933, and moved into a flat in North Wembley not far from Ethel's mother's house. I have already mentioned that Barnato had offered us a cottage in the grounds of his house near Lingfield, but before we could take him up on the offer there was a terrific fire at Arden Run and the house was completely gutted. Barnato himself didn't rebuild the place, and moved permanently into London, so that put an end to our plans for a life in the country. It also meant that we had to move the entire racing car preparation and building work to London, at garages in Belgrave Mews West, where Barnato already kept his town cars and his chauffeurs. Not that Ethel minded at all, because it meant that my commuting journeys to and from North Wembley and the new workshop were only a few miles and even in the London of 1933 that was quite a relief. There may have been only a million-and-a-quarter cars on our roads, but they were all a lot slower than now, and most of them seemed to be around me in London!

It was during 1933 that Barnato, who hadn't raced himself for at least two years, decided that he wanted to make a unique Brooklands car to have a go at Mountain Circuit racing in particular. He had his own ideas about the car, and because he was my boss I couldn't really argue with him. Whatever I may have felt about the merits of his schemes, Barnato wanted a Special with twin rear *and* twin front wheels, thinking that this would give us extremely good braking and traction. It had to be light and incorporate the most powerful braking system I could obtain. I tried to talk him out of the twin front wheels, at least, but at the time he wasn't interested in compromise, and I had to think hard about this when I started the design.

After the fire at Arden Run, Barnato moved into his sumptuous flat in Grosvenor Square, opposite the spot where the American Embassy now stands. His premises in Belgrave Mews West, just round the corner from Belgrave Square where the RAC Motor Sport Division is housed nowadays, consisted of just a big double garage with a large flat above where Cyril de Heaune (first chauffeur) and his wife and the second chauffeur used to live. Barnato thought there was enough room for me to build the new car there, even if I had to share with his road cars. I didn't really think so, and I knew the chauffeurs were not very happy about the arrangement, but we hit it off very well together. As it happened I only used the Mews for about two years, because once the Special was completed we moved it out to Brooklands where it lived permanently in a corner of the Thomson and Taylor garage inside the track.

As an aside, one of my recollections about the Mews is that our garage was opposite those of the Duke of Bedford - that would be the 11th Duke, of course. He was a rather peculiar man, and he never brought his Woburn cars into London, nor ever let his London cars out of London! We discovered that he always arranged for his London drivers to meet the Rolls-Royce from Woburn on the outskirts of town, where the occupants would transfer to the London-based car for the rest of the journey. He never wanted his Woburn Rolls-Royce to be driven in London in case it was damaged, while the cars he kept in London were old, though beautifully kept.

By the time we got down to building the Special, which the Captain had kindly allowed me to call the Barnato-Hassan Special, I was well used to conditions in the Mews, for we had already used it temporarily to build up the special 8-litre in which Clive Dunfee had been killed in 1932. I knew that Barnato would not be driving the Barnato-Hassan himself, because Dunfee's accident had finally put him off Brooklands driving. He was much more of a road-circuit enthusiast, and was probably the best, or one of the best, in the world at that time.

I finally managed to persuade him to drop the idea of having twin wheels on each axle - to my great relief - and I started to design the car round a $6\frac{1}{2}$-litre Bentley engine which we had standing on the floor at the time, our only 8-litre being installed in the other rebuilt car with its coupé body. People have often asked me just how much science and theory went into designing the Special. I never sat down to work out chassis stresses and suspension layouts on paper but instead I designed the main chassis side rails so that Bentley front and rear axles would fit, and a lot of other standard Bentley components could also be used. The chassis was a lot narrower than the standard one, because we only intended the car to be a single-seater. The side rails were more or less to the standard Bentley shape at the front, then I sloped them down to pass underneath the axle tube. At Barnato's insistence I fitted an Armstrong-Siddeley Wilson-type self-change gearbox, though bitter experience soon told us that it would get too hot, and it was only used for one or two races, after which it was replaced by a proper Bentley 'crash' box.

Barnato wanted the car to have outstanding brakes - he was still intent on using it on the Brooklands Mountain Circuit where braking was very important - so I rummaged around and eventually found that hydraulically operated Lockheed brakes intended for commercial vehicles would fit, and I linked these to a standard Marelli servo motor.

Of course, I couldn't make up a body-shell, but in the 1930s there were plenty of little specialist companies to tackle such things, and the first shell we fitted had an all-enveloping smooth shape around the mechanical components (though the wheels were exposed), with most of the tail occupied by a very large petrol tank. The body was tall because of the great height of the Bentley engine, but not all that narrow as we wanted a smooth shape without projections, which in this case meant covering the carburettors and the magnetos, which seemed to stick a long way out. Even the exhaust silencer was enclosed!

I spent a lot of time tuning the 8-litre engine which we used after the first race. There still wasn't the racing experience on the 8-litre that we had amassed on the 6½-litre Speed Six, though the two were very similar. I had to have special pistons made to raise the compression ratio so that we could use alcohol-based fuel, and we experienced quite a lot of trouble with connecting rods breaking in that engine. It wasn't until a couple of years had passed that I designed special tubular rods which cured the trouble, and as far as I know, the car, owned by Keith Schellenberg since the 1950s, still runs in this condition.

The Brooklands Engineering people did most of the special machining I needed for the car. They had sheds behind the paddock at the track, and they were really astonishing. One of their specialities was in producing pistons quickly, and normally one could go along to them, sketch out the sort of pistons wanted on the back of a Players cigarette packet, and pick up the set within a week, or possibly two weeks if the pistons were very special, say for an alcohol-burning Bentley. I suppose there were no more than half-a-dozen people working on lathes, and I think the secret was a set of patterns for casting which could produce piston blanks almost Meccano-fashion. They had a standard design, a very strong one with conical webs running down the centre of the crown to the gudgeon pin bosses, and they could make up pistons to any bore wanted. Nowadays . . . well, that was just one of the problems which became more and more serious when we were designing engines at Coventry Climax much later.

I really have no idea what the Barnato-Hassan cost to build, but one could get work done very easily, and for such good value in those days (I suppose a big bill would be for £25). Many of the parts were standard Bentley bits, which I could collect from the Kingsbury works. I don't suppose the whole car cost £1,000 to build, even allowing for the special chassis members and the bodywork done by a tiny firm in Camden Town, though I don't think Captain Barnato would have quibbled too much if the car had cost a lot more, because he was determined to have the fastest car at Brooklands. Later, however, even he got fed up with seeing the car use up a complete set of Dunlop track tyres at every meeting!

We finished the Special during 1934, by which time its purpose had changed from Mountain Circuit use to being a very fast Outer-Circuit car; in fact the car was never raced on the Mountain Circuit. It was finished nearly a year after John Cobb's Napier-Railton, with its 24-litre Napier Lion engine, had been unveiled. Barnato never had any ambition to build a car like that. He thought the Railton was a monster – we all thought that – and in a way we considered it was cheating because it was so vastly different from any other car

Oliver Bertram in the original Barnato-Hassan, a shot taken in the paddock at Brooklands

A famous Gordon Crosby charcoal sketch showing two of my Specials neck-and-neck at Brooklands; the Barnato, with Bertram driving, is the higher, with Bill Pacey below it and Brian Lewis' Lagonda behind. This was the 1936 Brooklands 500-Miles race, in which the Pacey finished second to Freddie Dixon's Riley and the Barnato retired with a broken connecting rod. (Reproduction courtesy of *Autocar*)

which raced at Brooklands at the time. Just as long as Cobb could keep it in tyres and keep it on the track it just had to take all the records, but it was interesting that Cobb's outright Brooklands lap record was set in 1935, and he never went faster after that. We knew that we had built a very fast car, and it made it all very satisfactory to be on equal terms, more or less, with such a huge-engined monster. Whereas we could drive the Barnato-Hassan flat-out, or nearly so, round most of the track, the Railton had to be driven more like a road car, on part throttle, on the bankings. You can confirm this from the much higher speeds the car set up when it was racing round a big circle on the salt flats at Utah, in America, on record attempts.

When I had finished building the car I put some temporary road equipment on it, including mud-splash wings, and I actually drove it from the Mews to the track at Brooklands. That created quite a stir, as it really was a dramatic-looking car; it was the only time during my association with it that the car was driven outside the confines of the Brooklands circuit.

Once it was installed at Thomson and Taylor's premises, my daily trip to work changed again, this time from North Wembley to Brooklands, which was a bit further but not nearly as busy. It was about the same time that Ethel's mother decided to convert her big house into flats (she intended to go on living in one of them) and so we moved into the ground-floor apartment, which just happened to have a lock-up garage in the orchard. That garage was going to be useful, later. My first son, John, was born in 1934, so all in all it was quite a significant year for the Hassan family.

From then, until I left Barnato's employ at the end of 1936, the whole of my working life was centred on Brooklands. It had a pleasant 'village' atmosphere, where everybody knew everybody else, and I never became bored with it because there always seemed to be so much to do. The circuit itself didn't interest me much, except as an excellent proving ground, and I wasn't particularly absorbed by the racing, except as a proof or otherwise of my work. Even when I went to Le Mans with Bentley I spent most of my time with my back to the track, and it was only when one of our cars was coming into the pits, and I felt I could do something to influence the result, that my interest would perk up. I was much more of an engineer and designer than a racing enthusiast. I could drive the car quickly in testing and in practice, but I never drove it in a race. I was quite happy to leave all that to Bertram and the others.

The Barnato-Hassan was a very fast car indeed. After I had made the engine reliable and had tuned it properly, Oliver Bertram took the car round Brooklands at 142.6 mph, which gave us the lap record for a short time. Later, with a slimmer body, he lapped at 143.2 mph, which was only a whisker under the speed finally put up by John Cobb's monster. Bertram, a barrister, was a very dashing young man, who was already well-known at Brooklands for the way he drove the ex-Cobb $10\frac{1}{2}$-litre Delage before Barnato asked him to drive the Barnato-Hassan. Like a lot of the well-to-do chaps who raced at Brooklands, he was always surrounded by smart and pretty girls.

It really was very frustrating trying to win races with the Barnato-Hassan, as we spent most of our time trying to beat the handicappers rather than the other cars. Every race at Brooklands was run on a handicap basis, which meant that a car's previous performance was taken into account, and Ebblewhite and his merry men tried to handicap every race so that there would be a mass dead-heat; a monumental crash would have ensued if they had been successful! The problem with the Barnato-Hassan was that it improved very little from the first time we put the 8-litre engine into it, which meant that we never had a new 'secret weapon' with which to surprise everyone. Barnato soon got tired of this sort of frustration, and after he withdrew the car from Brooklands (soon after I had left him) he never went racing again. I think he was convinced that the car with the best performance should win races, and I must say I agreed with that.

My next Special, which became known as the Pacey-Hassan, was built with that handicapping system in mind, and it worked well at the track for about a year. Bill Pacey wanted to race his old sports Bentley with much more success than he had managed to achieve, and because of my reputation with the Barnato-Hassan he wrote to Captain Barnato to ask if he could get me to tune his car in my spare time. Naturally, Barnato agreed to this, so I suddenly had two cars to look after, one during the day and one in the evenings and spare weekends. Once Pacey decided that he wanted to go a lot faster we persuaded him to have a special track car built, and we carried out the construction in the garage in the orchard in Wembley. Most of the building work was by my brother-in-law, Wally Saunders, who had dragged me out of the crashed Bentley at Montlhéry in 1926. Wally was still working in the Bentley service workshops, so the new car was a spare-time project for both of us.

I set out to design it according to the same principles as the Barnato-Hassan, though it was a lot lighter by the time we had finished with it. It was altogether a much simpler car, and I seem to remember that it cost only about £600 to build. Not long ago I heard that it had been sold again, but this time for a reputed £25,000. How times change! Despite Brooklands being a motor racing 'village', Wally and I kept ourselves to ourselves, and in this way we produced a secret weapon for improving the car from race to race.

The $4\frac{1}{2}$-litre engine, like all Bentley designs apart from the final 4-litre, had a fixed cylinder-head, in unit with the block, and the compression ratio was varied by differences in the depth of the block. So we built the $4\frac{1}{2}$-litre racing engine with a whole stack of compression plates under the block, on top of the crankcase, having already machined the base of the block considerably. This meant that we started the season with a very normal compression ratio of about 6.5 to 1. We won races at once, of course, and were re-handicapped, but every time that happened I felt that we needed to rebuild the engine. Now I wasn't going to admit to being a forgetful mechanic, but somehow we always managed to complete the rebuild with fewer of these gasket plates; it really was very careless of me! Of course, as the compression ratio went up we changed the fuel mixture to suit, and the car kept on winning races. But that only lasted

Bertram again in the original Barnato-Hassan. There was scope for narrowing the body still further and improving the shape. The track tyres had virtually smooth treads even when new

The slimmed-down Barnato in 1936 form, when it was a strong challenger to John Cobb's 27-litre Napier-Railton. Bertram is in the car with Woolf and Jacqueline Barnato alongside. On reflection, I wish I had changed my overalls for this shot

Bill Pacey lapping at 129 mph in the Pacey-Hassan. The notorious Brooklands bumps have taken all four wheels off the ground

until we ran out of plates to 'forget', by which time the compression had risen to around 12 to 1! It gave Pacey a full year with a steadily improving car; he won the Brooklands Gold Star and took a place at every meeting he contested that year.

By this time Ethel had come to the conclusion that I was taking root at Brooklands, and she decided that we would have to move to a house nearer the track, so we rented a bungalow in Shepperton after the birth of our second son, Peter.

For 1937 Wally Saunders and I decided that we should re-engine the Pacey-Hassan Special, so that (we hoped) it could start winning all over again. We decided to use a 4½-litre crankshaft married to a bored-out 3-litre cylinder block, a 'blower 4½-litre' bottom end and a Zoller supercharger. We used 'blower' 4½-litre rods and the latest crank and it all hung together for the season, but Pacey had little success with it that year.

Building and tuning exciting Specials like this presented quite a contrast to the cars I used for everyday transport. When I started to work for Barnato I had a Morris Minor, one of the original overhead-cam models whose engine went into the MG Midgets in modified form. That was a very simple little car, and one I used for thousands of miles. Even repairs were very simple. I was once coming back from Devon when I noticed that the oil pressure was dropping to the danger level. I knew that the bearings tended to wear out rather rapidly on that engine, so I stopped, dropped the sump at the side of the road (keeping all the oil instead of draining it), took the big-ends down one by one, sorted them out until they were tighter again, assembled the whole thing and then drove home with the oil pressure restored to normal! Later I sold the Minor to Stan Ivermee, who covered even more miles in the car than I had. Just before I was married I replaced it with a Riley Nine from Thomson and Taylor, who were the agents. I raised the compression ratio and fitted twin carburettors, which no doubt contributed to the three broken crankshafts that I suffered with the car! It lasted me for a couple of years, but after it had broken down for the last time I was very lucky to be given a nice old 'Bullnose' Morris Cowley by McKenzie, who later became famous for his work on Forrest Lycett's phenomenal old 8-litre Bentley. I was highly delighted to drive that car away, especially as it was free, and I used it for quite a time. But of course it was getting old, and old age and the hard springing combined to shake the radiator to pieces on my 40-mile daily runs from Wembley to Brooklands and back, so eventually I ditched the 'Bullnose' radiator and replaced it with a square radiator and bonnet which I scrounged from Thomson and Taylor. As I had to keep the round-radiator scuttle it all had to be tied together with string, and in the end it was so ragged and scruffy that I was stopped by a policeman in Hammersmith Broadway and more or less warned off! I replaced it with yet another cheap car – a drophead coupé Alvis with a rather droopy hood which cost me the princely sum of £10.

Private motoring wasn't costing me much money, but then I didn't have much. The cars I bought were so cheap that by the time they packed up I just ditched them and started again. In fact it was the work we did for Bill Pacey, and his pleasure in the results achieved, that really helped me to get my first reasonable car. Pacey was in the retail motor trade, with premises in Golders Green. He supplied me with a year-old Austin Ten for £110, which was big money for me, and a far cry from that first GN at £4.50!

I was to become involved in the design of yet another Special in 1938, but much was to happen before then. One day during 1936 Captain Barnato asked me to call round and see him at his Grosvenor Square flat, and as I often went up there to talk about details of the new cars with him I had no idea that this meeting would be different. No sooner had I settled down than he said that he had decided that I was now wasting my time with him, especially as it looked as if the Barnato-Hassan couldn't win races because of the unfair handicapping arrangements. He said he thought it was time I took on further responsibilities, and that I should seek for opportunities in industry.

Ever since I had first known Barnato at Bentley Motors he had been very kind and friendly towards me. Although he was always very much the boss, I like to think we had much more than a normal employee-master relationship. He was by no means interested only in motor sport and the industry; for example he was a very fine cricketer and a theatrical 'angel', as well as the many-faced businessman with company brass plates down both sides of his office door in London. With all this activity he was finding less and less time for motor racing. Explaining his decision to me he said, "I have not driven a race car since Clive Dunfee was killed in the 8-litre Bentley Special, and I think you should take the chance of moving on before I drop out of the sport altogether".

He told me that he had many contacts in industry, and he very kindly offered to set up interviews with important engineers so that I might find a new job without too much searching. There was no question of him getting rid of me, but there was no doubt that he was beginning to lose heart due to the Barnato's handicapping problems, and it was very good of him to take the trouble to point me in the right direction. He said that as he was still on the Board of Directors of Rolls-Royce, he wanted me to go up to Derby to have a talk with their technical chief, Mr W.A. Robotham, with a view to working there.

This was all rather sudden, and of course somewhat disturbing at the time. We had only recently moved to the Brooklands area, and I was very happy in my development tasks based on the Thomson and Taylor workshops. Now, of course, I can thank the Brooklands handicappers for bringing about such a sudden and important change in my life, but at that time I had no idea where it was to lead. It looked as though the first part of my life, in motor racing, was at a close, and that the next stage, in industry, was about to begin.

Barnato-Hassan and Pacey-Hassan Specials

Building the Barnato-Hassan Special took the best part of a year, from autumn 1933 to well into 1934. The first body-shell, made in Camden Town, only provided for a single occupant, but it was effectively a one-and-a-half-seater since the driving position was offset so that the maximum number of unaltered Bentley components could be used. The nose of the shell incorporated an oval cowl which enclosed the tall radiator and the front dumb irons, while the tail was almost entirely fuel tank. Hassan never knew the power output of the engines he fitted, as he never had access to a test bed, nor the time or inclination to find out. After all, he said, the times around the Brooklands circuit were all-important, and it didn't matter what power output was obtained as long as the results were achieved. Both the engine and the transmission came out of Barnato's old Speed Six, after early trials with a Wilson-type preselector box had shown the latter to be unsuitable for this particular car.

Once the car was completed, and Hassan had fitted it with a set of temporary wooden mudguards, and driven it down to Brooklands, where it lived for the next few years, it was never raced anywhere else. Like many other well-loved Brooklands Specials, it was quite unsuited to any other venue (apart, perhaps, from straight-line sprints at Brighton and elsewhere), and certainly could not have been raced at Donington, which was Britain's only other permanent racing track. Yet surprisingly, in recent years Keith Schellenberg has campaigned the car in a variety of Vintage Sports Car Club events at Silverstone and Oulton Park – localities its designers could never have envisaged when it was new. It now lives in honourable retirement in Schellenberg's castle in Aberdeenshire.

The car was finished in time for the 500-Miles race at Brooklands in September, 1934, when it was driven for Barnato by Dudley Froy and Earl Howe. In practice it gave trouble with an out-of-balance propeller shaft, and it had no chance to shine in the race before the hard-worked old 6½-litre engine broke a connecting rod and holed the crankcase on each side.

During the winter of 1934/35 it was housed in a corner of the Thomson and Taylor premises, along with many other famous and infamous Brooklands Specials. Hassan's main job was to install the race-tuned 8-litre engine, complete with triple SU carburettors, fed through a long ram-pipe mounted along the bonnet on the offside of the car. Though Woolf Barnato drove the car from time to time on tests, he never actually raced it himself, for he had given up active competition some time before the car was completed. Instead he decided to play the part of wealthy patron to up-and-coming drivers, and for the Barnato-Hassan he chose

Oliver Bertram, the young London barrister. When Bertram made his race debut with the car at Easter 1935, the Brooklands handicappers did not have their true measure for the car's brief appearance in the 500-Miles race the previous season had been unencouraging, and despite starting on scratch, the Barnato-Hassan won the Easter Senior Short Handicap, with a best lap speed of 134.97 mph – not bad for a relatively undeveloped car! The Barnato's problem thereafter was that it attracted attention from the handicappers, and for the rest of its Brooklands life Hassan was fighting a losing battle against their mathematics. Perhaps this is an appropriate time to discuss the system.

For many years the BARC (Brooklands Automobile Racing Club at that time) had realised that there were many more racing enthusiasts than wealthy men, many more small cars than fast cars, and little opportunity to make the Brooklands circuit equally difficult for all. Therefore, most Brooklands races – especially those held on the Outer Circuit – were organised on a handicap basis, either by starting the cars at intervals in short races, or starting everyone off together in the long races but allocating a complex number of credit laps which were taken into account at various points in the race. The object was to ensure very close finishes to all the short races, and the handicaps were devised by a diligent timekeeper – Mr Ebblewhite – whose profession as a musical shop owner was at variance with the excitements of Brooklands.

Handicaps were worked out on the basis of a car's previous Brooklands performances (along with those of its driver), therefore a winning car, driven flat-out, would soon be handicapped back into the ruck for subsequent races. The only way a car could continue to win was for it to be 'pulled' at first (a practice frowned upon as much at Brooklands as it was by the Jockey Club, for similar laudable reasons), or for its engine tune to be improved regularly until the edge of mechanical disaster was reached! Inevitably, the Brooklands system, which was universally hated by the owners and drivers of very fast cars like the Barnato-Hassan, the single-seat 'blower' Bentley and the Napier-Railton, was to frustrate Woolf Barnato's enterprise with this splendid car for the next four seasons.

Handicaps were so speedily revised by Ebblewhite that between winning its first race and coming out for its second on the same day (for the Long Handicap) the Barnato was asked to 'owe 15 seconds' instead of merely starting from the scratch mark. This was too much for the brave young Bertram, who was unplaced in that event, but the significant point about the second race was that the Barnato-Hassan's lap speed was worked up to a remarkable 137.96 mph, which would have been good enough for an outright lap record only a couple of years earlier, although this was now firmly in the hands of John Cobb at 140.93 mph with the 24-litre Napier Lion-engined Railton, which was being worked up to its full potential. Incidentally, at this sort of speed the big cars were by no means flat-out all the way around the rim of the Brooklands bankings. Tyre limitations and the track's notorious bumps meant that even the bravest drivers would have to lift off at times, and Bertram's near-138 mph lap speed meant an instant speed of around 150 mph on the straights. Those of us who have seen the reverse curve, however *slight, at the Fork near the Vickers Sheds may marvel!*

By Whitsuntide, Hassan had squeezed a little more speed out of the green-painted Special by removing the front brakes, 'discing' the rear wheels to cut down on turbulence, and trimming the weight a little (although it still weighed 29 cwt), and later in the year Bertram managed to lap at 142.70 mph. This was better than Cobb's outright figures, but it could not count as an official best figure as these always had to be set by making an individual time trial. Even so the car's handicap left it unplaced in the Short Handicap, and Barnato withdrew from the Long Handicap in some pique. However, later in the day Bertram brought the car out for a lap-record attempt, and almost equalled his earlier figure with a sparkling performance at 142.60 mph to claim a new all-comers' record, although it was not to last for long; later that year Cobb took out the Napier-Railton and added a further 0.84 mph to the record, a performance that was not to be beaten during the remaining four years of racing at Brooklands. Of course one could expect the relatively unstressed 24-litre Napier-Railton to go so quickly (it had regularly achieved this pace for up to 24 hours on other tracks) but Bertram's performance was in an 8-litre car which owed much of its ancestry to one of the most luxurious saloons ever designed.

Later in the year both the cars were entered in the BRDC's 500-Miles race, where they fought a gallant battle (on handicap, naturally) for the lead, but the Barnato-Hassan was eventually forced to retire with a split fuel tank and the Railton went on to win.

During the winter of 1935/36 Barnato requested Hassan to reduce the frontal area as much as possible in order to improve penetration. The big problem was the 8-litre Bentley engine which, though slim except for the carburettors, was decidedly high, but the revised Special, which was ready for the 1936 season, had a very slim shell and tiny air intake with bulging body sides over the chassis frame. The driving position was moved to the centre and lowered, with raked steering, and the tail now incorporated a headrest and a new fairing over the big fuel tank. But unfortunately, Hassan also tried to raise the power output of the already sharply-tickled engine, and in 1936 there was to be more than one occurrence of broken connecting rods (particularly disastrous because this could mean damage to the very rare 8-litre blocks, which were cast in electron) while the car was to prove no faster than before. Breakages, usually near the small end, and not completely cured by fitting special lightweight pistons, put the car out of the 500-Miles race and 'The Star' Gold Trophy race.

Barnato by now was so depressed that after Hassan had left him the car languished for a time. In 1937 it appeared only once, when it actually defeated the handicappers by winning the Second October Handicap with a best lap speed of 137.68 mph. Hassan, meantime, had returned to Brooklands and Thomson and Taylor, and found time to have some special tubular connecting rods made which cured that problem for good. The engine now ran on a special ethyl/benzole/alcohol mixture with a compression ratio of 8.7 to 1 and felt rather quicker than before. However, Brooklands seemed to have exacted its own limits on big-car speeds, and in spite of the new-found reliability and performance Bertram was only able to edge up his previous achievement by fractions. His best lap speed in 1938 crept up to 143.11 mph, only 0.33 mph behind the big Railton, and the best-ever Brooklands performance by a car fitted with a car engine. This, however, was the swansong of the Barnato-Hassan Special, which was not to be raced again before the war.

The later Pacey-Hassan, built as an 'evenings and weekends' Special in the garage in North Wembley, while Hassan was still employed on preparation and maintenance of the Barnato-Hassan Special, used basic Bentley mechanicals in a new chassis frame, with a lightweight body wrapped as closely around the tall engine as possible.

Because they were not financed by the Barnato millions in the building of this new car, the two Wallys had even less in the way of mechanical aids, but the Pacey, which was started in 1935, was ready to race early in 1936. The Hassan-designed chassis frame (laid out according to the same rule-of-thumb precepts as that of the Barnato car) was made by Rubery Owen, but assembly and lightening (with all holes drilled laboriously with a hand brace) were completed in the garage. Pacey's old Bentley sports car was stripped out and the $4\frac{1}{2}$-litre engine was over-bored by 1 mm (to 101 mm) to give a capacity of 4,487 cc. Hassan's habitual tuning tricks were applied internally, including the use of Martlett pistons, as in the Barnato, stronger connecting rods, and the special 'handicap-beating' stack of compression plates. The sleek and narrow body-shell, which bore a family resemblance to the final Barnato, was evolved by the Gray brothers at Thomson and Taylor. There were no front brakes as this was to be strictly an outer-circuit Special. Like that of its big relation, the Pacey's streamlining was hampered by the very tall engine - as tall as the 8-litre of the Barnato, of course.

Hassan purposely began with a fairly mild degree of tune, and by using the former engine and chassis numbers of the old sports car ensured a good handicap to start with; the engine had a compression ratio of only 6.5 to 1 with all its compression plates in position. The car first appeared at the Brooklands 1936 Easter meeting, when Hassan's track-craft was proved almost at once. In its first race, a typical 'Short Handicap', the Pacey-Hassan lapped at 117.46 mph and won its race by more than three seconds - a lot for those Ebblewhite-dominated days. Promptly, as with the Barnato, the handicap was revised and on the same day Pacey was unable to win another race despite lapping at nearly 120 mph.

Some of the compression plates had to come out to allow the combined block/head to sit down a little further around the pistons and by Whitsun Pacey was able to lap in the Gold Trophy Handicap at no less than 128.03 mph and win from John Cobb's V-12 Sunbeam and Marker's $6\frac{1}{2}$-litre Bentley. By the end of the summer the car was back on the 'scratch' mark among far bigger opposition, and even a lap speed of 128.36 mph was insufficient to finish better than second to 'Goldie' Gardner's MG in a handicap race. Later in the day the Pacey lapped at 129.03 mph - its best Brooklands performance - but it was all in vain. At the close of 1936, when Hassan had gone off to work for ERA at Bourne, the car finished in a very creditable second place in the BRDC '500', averaging nearly 116 mph for more than four hours, including pit stops.

For 1937 one may assume that all the compression plates had been removed – which meant a compression ratio exceeding 10 to 1 on 'dope' fuel – and the performance was at its height. To win races it was necessary to do something drastic about the engine, and the Hassan-Saunders partnership settled on a venerable 3-litre Bentley, this time over-bored by 2 mm to give 2,956 cc, and modified to include a 'Blower' 4½-litre crankcase, 'Blower' crank and rods, and of course a Zoller supercharger. The conversion wasn't completed until the middle of 1937, and the car only raced in the 500-Kilometres race. The power output of the hybrid engine was not known, but tests showed it to be less powerful than the original 4½-litre. It was therefore a pleasant surprise to find that Pacey, on the large balloon-section track tyres, could lap at 120 mph. However, success in this '500' was to be harder to achieve than before. The Brooklands bankings had become even rougher over the years, and the Pacey-Hassan did not enjoy the latest bumps. With a '500' measured in miles rather than in kilometres, as it was in 1937, the car would undoubtedly have been withdrawn; Pacey, however, persevered and eventually finished a creditable eighth at 108.9 mph. However, after this showing Pacey, like Barnato, decided that there was little enjoyment to be gained in trying to outwit the handicappers and scrutineers, and he retired the Pacey-Hassan forthwith. It did not race again in the 1930s, but has since been restored to its original glory and is seen from time to time at VSCC events.

The car itself might indeed be original, but its value most certainly is not. Built at a price in the region of £600 in 1935/36, it was to change hands over a quarter of a century later for a reputed £25,000!

3
Records and radials

When I was given the opportunity of working for Rolls-Royce at Derby, their reputation was as high as it had ever been. Like the old Bentleys, their prices were very high, yet there was still a long queue to buy. Until then a Rolls had never excelled in pure speed but they were very good in engineering refinement, while in sheer quality of finish they were untouchable. The structures were excellently designed and used the best-quality materials, and they were more carefully assembled than any other make. No-one doubted that they were the 'Best Car in the World'.

In the early 1930s performance had been no great shakes, but the V-12 Phantom III had done much to change this. Of course, when the new 'Bentley' came along I was very interested to see how much of the old designs had been used, but like many enthusiasts I was disappointed to note that there was not a single part common with the 8-litre or even the 4-litre cars I knew so well, except perhaps in radiator shape and in the use of the same winged badge.

Rolls-Royce and Rolls-Bentley roadholding was extremely satisfactory, and when I went up to Derby to keep an appointment with Mr Robotham I found that they were striving to improve the cars by introducing independent front suspension, and in fact the job he was offering me was as a development engineer on this suspension. Not many cars had independently sprung front wheels then, and those which had were not always very effective; almost everything from a cheap Morris to a Rolls-Royce relied on half-elliptic leaf springs and a beam axle. With the tyres we had and the bumpy road surfaces which were common it was quite normal for cars to suffer from wheel-wobble, tramp and 'shimmy' - something that no amount of careful tuning of springs and dampers could cure. It didn't help that by modern standards shock absorbers were really very crude, and of course the chassis frames were none too stiff at the front either.

The job that Robotham offered me was really quite a good one, but I turned it down, even though Barnato had done me a considerable favour by arranging the meeting in the first place. I suppose it didn't help that it was a perfectly miserable day when I travelled to Derby, and since Derby wasn't a very attractive place at the best of times that put me off. Also, I didn't like the way they intended to pay me. The basic rate was rather less than Barnato was already paying me, but Robotham said that I should not worry as there was always a very large annual bonus; the previous year it had been 100 per cent, and he saw no reason why it should be reduced in future. All the same, I didn't like the idea. There was no question of my not getting on with Robotham, and I suppose the clincher was that I just didn't like Derby.

I had to go back to Captain Barnato and tell him that I had turned down Rolls-Royce, but he was very nice about it, and after quite a bit of discussion he decided that I should talk to Humphrey Cook, who lived just round the corner from him. Cook was one of the original financiers of the ERA racing car firm, as well as having been a member of the Bentley racing team as a private owner, and since I knew many of the people connected with ERA this sounded much more attractive. So I talked to Mr Cook, who suggested that I had a chat with Peter Berthon, who was ERA's Chief Engineer at Bourne. The idea of working up in the wilds of Lincolnshire was a bit of a snag, particularly as we had just settled near Brooklands, but since ERA was a compact little firm making very interesting products I didn't worry too much about it. Berthon offered me a job running the assembly shop where the latest series of ERA racing cars with Porsche-type independent front suspension were to be built.

This sounded very interesting, but I was still cautious about a permanent move up to Bourne, so although I accepted the job I decided to live in digs until I saw how things turned out. This meant that I was only home at weekends, and since I was still interested in cars at Brooklands (the Barnato and the Pacey, in particular) I didn't get an awful lot of home life. I began to work at Bourne towards the end of 1936, but I was only to stay there for a matter of six months or so.

ERA was a very successful racing car by then, even though it was only a *voiturette* with a much-modified Riley engine. Not everyone realises that Reid Railton had designed the chassis for the car, or that my friends at Thomson and Taylor assembled the complete car apart from the engine and gearbox, which were fitted at Bourne. So even though I would be a long way from home and what I now considered to be a permanent workplace at Brooklands, I would not be out of touch with the friends I had made there. In any case I would still be involved in Brooklands racing at weekends, for at that time Wally Saunders and I were modifying the Pacey-Hassan for its second Brooklands season.

The decision to make revised ERAs with independent front suspension had been taken before I went to Bourne, and it was obvious that the 'Auto Union-type' Porsche suspension offered a simple conversion without too much development. I think it was obvious to all of us who saw the cars racing at Brooklands and elsewhere that the combination of a solid front axle with cart springs and a whippy chassis frame had outlived its welcome and had to be replaced. To be fair to Railton, who designed the original chassis, there was no tradition of independent suspension at all in 1933 – all the successful Grand

Prix cars like the Alfas, Bugattis and Maseratis had solid axles, and those early ERAs were very successful, too. However, 1934 was the year when Mercedes-Benz and Auto-Union came along with radical ideas to change everyone's standards! Although the ERAs did not race against the big German cars, Peter Berthon had ample opportunity to study their performance as his cars often raced in subsidiary events at the same circuits at the same meetings. In 1936 Auto-Union dominated Grand Prix racing, and Berthon thought that their Porsche-type suspension was a more promising layout than that of the Mercedes-Benz; as far as I know the ERA system was actually designed in Dr. Porsche's office.

I arrived at Bourne towards the end of 1936 in time to supervise the building of the first car. It wasn't simply a case of converting existing structures as the entire chassis was new, and this time we tackled the job ourselves instead of farming it out to Thomson and Taylor.

The new design was box-sectioned with strong tubular cross-members, and was very strong in torsion, whereas the original design was so whippy that we reckoned that it looked after the front suspension without the need for any springs!

All the time I worked at Bourne I stayed in some digs next to the railway level-crossing and drove home to Wembley at weekends. My little Austin Ten was really having to work hard for its keep because I had a 200-mile round trip every weekend. Ethel had given up the rented bungalow in Shepperton, and was staying with her mother in Wembley, which was convenient as the Pacey-Hassan Special was being modified in the garage in the orchard. I must say everyone was very good about the clutter this used to bring into the house – for instance the main chassis members for the Pacey were delivered by Carter Paterson and had to live in the hall until I got around to moving them out!

Early in 1937 I heard from Ken Taylor of Thomson and Taylor that Reid Railton would like me to return to Brooklands to work with him on a new project. I knew from what Ken told me that the job was in connection with the new Land Speed Record car which Reid Railton was designing for John Cobb. I had been feeling the pinch financially, what with living in digs and having to maintain my family in London as well, so it all sounded quite fascinating, and it fitted in well with my future hopes and plans. I duly made my peace with ERA.

Railton himself had been Chief Engineer at Thomson and Taylor for a number of years, and was already famous for his work on the later Campbell Bluebirds and on John Cobb's Napier-Railton which had proved to be just a fraction faster than my Barnato-Hassan Special round the Outer Circuit at Brooklands. I saw Railton one weekend, and he explained that John Cobb had commissioned Thomson and Taylor to build an out-and-out Land Speed Record contender with a target speed of 400 mph! Railton was absolutely confident that this could be achieved, and whereas that sort of talk from most other people would be considered very brave, his reputation was so high that none of us doubted his ability to design such a car. In 1937 the record still stood at 301 mph (Sir Malcolm Campbell's Bluebird) but George Eyston had built a real monster of a car in an attempt to beat it that year.

Railton said that he wanted someone to be in charge of the search for suitable companies to make components, and to progress these through from designs to the actual hardware. I didn't really have a title – it wasn't thought important at the time – but I suppose I was a sort of sub-contracts manager. There was never any real difficulty in getting well-known firms to tackle jobs for the new car, although the engineering industry was brisk owing to the threat of an impending war; Railton *and* Thomson and Taylor were both famous and well-respected and making parts for a Land Speed Record car was thought to be good for prestige and publicity.

Railton showed me his design for the new car, which looked exciting, beautifully engineered and much lighter than its rivals. I didn't really hesitate about his offer because everything about it appealed to me. I would be moving back to Brooklands, and would be able to live at home all the time. Although I had made good friends at Bourne, I left with few regrets, and instantly found myself back in the friendly village-like atmosphere of Brooklands. It also made it easier to keep up with the fortunes of the Barnato-Hassan and Pacey-Hassan cars.

Until the Railton was built, the general concept of a Land Speed Record car was of a monster with aero-engines, probably not attractively shaped, and likely to be heavy enough to ensure adequate traction and stability at such speeds. It was a tradition which had evolved without development since the 1920s. Reid Railton's car was a remarkable design in every way. Its shape was aerodynamically very good, and although there were two aero-engines under the streamlined skin, the car's total weight was only about three tons – less than half that of George Eyston's Thunderbolt, which was nearly finished when the Railton project was started. Railton was out to save every possible ounce of weight, and in a way it was a very simple car. By accepted standards it was also quite small – about 27 ft long, and with a widest track (the front) of only 5 ft 6 in – yet the two supercharged Napier Lion engines gave a capacity of almost 48 litres and an output of around 2,500 bhp! The front engine drove the rear wheels, and the rear engine the front wheels.

Railton himself had such a reputation that I don't think many people doubted his absolute confidence that the car would succeed immediately. Although I was not involved in the design, I had to take responsibility for component manufacture by dozens of outside suppliers, negotiating delivery dates and costs, and – if it didn't all happen when it was supposed to – taking the sharp edge of Reid Railton's tongue. Of course, Railton was a very gifted designer, but he had a terrific ego and I never struck up a very close relationship with him. He rarely discussed any point of his design with anyone, and we thought him a rather snooty character! He once said in a magazine feature about the Railton that he expected his aerodynamic theories regarding the car to be vindicated by the appearance of depressions in the thin aluminium skin near the tail and when the time came he was right – the dents appeared, absolutely as predicted. In fact, there was very little wrong with the car, which took the record surprisingly quickly. We sent it out to the Utah Salt Flats with a big fin, but Cobb took the record without it. I do remember that there was a fair amount of trouble in dealing with the torque reaction from the Napier engine on the very narrow-tracked rear axle, which caused one wheel to lift, but even this was eventually solved with a very simple face-cam-and-links system. Apart from its shape and relatively small size, we mustn't forget that it was the first Land Speed Record car with four-wheel-drive and the first in which the driver was seated ahead of the front wheels.

It was always understood that when the car was completed I would stay at home and manage the service and racing-car preparation side of things while Ken Taylor and his mechanics, who had built the car, took it over to America for the record attempts. The Railton duly took the Land Speed Record at Utah at 350 mph, but then George Eyston, who was there at the same time with his Thunderbolt, went out the following day and beat John Cobb's time! Of course it would have been nice to go out to America with the Railton, but I must say I found the Brooklands-based work so interesting that I was not at all disappointed. I was very keen on my job, but even so I couldn't possibly have devoted as much time to it as Taylor seemed to. He lived in a bungalow adjacent to the works, but he seemed to begrudge even going to bed at night, and spent most of his time in the shops. He was President and a founder member of the British Racing Mechanics Club, an organisation that has continued to go from strength to strength ever since.

We were Alfa Romeo agents, amongst other things, at Thomson and Taylor, which meant that we had to service all the road cars and the latest racing cars which filtered over from Italy from time to time. There were a lot of fast and interesting Alfas at Brooklands, including the twin-engined Bimotore which Austin Dobson tried to master. Interestingly enough, the only car which got close to breaking the outright times put up by John Cobb's Napier-Railton and the Barnato-Hassan Special was Chris Staniland's Multi-Union in 1939. This looked rather like a Grand Prix Mercedes, and was the work of W.C. Devereux of High Duty Alloys Ltd, who incorporated many components from a racing Alfa Romeo including a 2.9-litre engine. Staniland didn't quite make it, but if racing had carried on into 1940 we were fairly certain that the car would have had the potential to take the record.

It was during this happy and interesting time at Brooklands that I became involved in the third and last Special to carry my name, one which was never as famous as the Barnato or Pacey cars, probably because it was neither as successful nor raced with such determination. The kernel of this car was that Dick and Pat Wilkins, well-off sons of a successful stockbroker, wanted to have a completely competitive car for Brooklands, and were interested in the Mountain Championship.

The Land Speed Record Railton outside Thomson and Taylor's workshops at Brooklands in 1938. Taylor, Railton, Cobb and Thomson are grouped around the cockpit and I am hidden away at the back, seventh from the right

The BHW Special, which I built for the Wilkins brothers and which used a straight-eight Bugatti engine. It was not as successful as the earlier Hassans

Reg Parnell driving the BHW in a Brooklands Mountain Handicap race at the 1939 Easter meeting. (Photograph from the William Boddy collection, reproduced courtesy Mr Boddy)

They had already acquired various racing cars, which were serviced by Thomson and Taylor, but had not been rewarded with much success. An old friend of mine from Bentley days, Jack Sopp, was also a good friend of the Wilkins brothers. Talking about future racing prospects in 1937 I suggested that if we were to build a new car, the performance of which was an unknown quantity to the handicappers, and if it was extremely fast but easy to handle, it might stand a good initial chance as the Pacey had done. Now Dick Wilkins was a good friend of mine, but although he was a good steady driver I knew he was not the world's fastest, so I looked him straight in the eye and said, "What we ought to do is build a car that is so fast that you can just drive it steadily round the corners!" Then we all had a good laugh, realised that it was possible, and started to design the BHW, or Bugatti-Hassan-Wilkins.

Along with Jack, we obtained a 4.9-litre supercharged Bugatti engine and designed a chassis and body-shell to accept it. I schemed a new gearbox complete with final drive, to operate along with a rear swing-axle layout based on that used by the Auto-Union of the period. That decision was actually a mistake, though it appeared at the time to be the ultimate solution. I persuaded another ex-Bentley friend of mine, 'Spud' Ayto, to carry out the actual design and detail of the car, and I made a lot of the drawings. We used Alvis independent front suspension and Lockheed hydraulic brakes, and also set to work on the engine, designing new connecting rods, pistons and other details, raising the compression ratio and making the engine more reliable.

All this took time, and it was not until 1938 that Dick Wilkins was able to race the car. The swing-axle at the rear made the car a trifle tricky to handle, but it was quite shatteringly fast. In its first race, Dick started in a Mountain Handicap, stalled the engine due to the rather fierce Bugatti clutch, but finished behind the winner exactly as far back as it had taken him to re-start. I never followed the car through, due to lack of time, nor had any hand in its maintenance, but it developed its own reputation, partly connected with the roadholding and partly with the performance. Reg Parnell bought the car for the 1939 season, and told me afterwards that he had had a lot of useful and rewarding racing with it. The car has also survived the last thirty-odd years, and still appears at VSCC meetings.

Life then was never dull, and I was finding it most pleasant working all the week on high-performance and racing cars, seeing how they performed at weekends, and living so conveniently close to work at Byfleet. However, just when I was beginning to feel settled again, and when Ethel could actually rely on seeing me during the week as well as at weekends, I received a very discreet approach regarding a new job – with SS!

I had had my first contacts with SS Cars of Coventry through the tuning and race-preparation which I had supervised in the workshops at Brooklands, and due to the kindness of my old friend Tommy Wisdom. SS had introduced a $3\frac{1}{2}$-litre version of their SS100 two-seater sports car in 1937. Although at the time I understand that William Lyons was not at all keen to see his products participating in motor sport, he and his Chief Engineer, Bill Heynes, were finally persuaded to consider race-tuning the engine. Tommy Wisdom, whom I had known for years as a successful racing and rally driver, was close to SS, and had won a race at Brooklands with a super-tuned engine which Bill Heynes had developed at Coventry.

The 3½-litre SS100 engine produced 125 bhp in standard form, which was enough to give the car a 100 mph maximum speed, but a lot more than that was wanted for short-distance sprint events and hill-climbs. The sprint engine had the compression ratio raised to something like 15 to 1, so as to run on neat methanol. The ports were opened up and polished, there were special pistons, polished con-rods, large SU carburettors – in fact the lot. Mr Heynes very kindly passed all this information on to me, as well as making all the special bits and pieces, in order that Thomson and Taylor could prepare a very special SS100 for Edgar Wadsworth to race in Ireland. The car performed quite well and the factory were very pleased.

Even so, it was quite a surprise when Tommy Wisdom approached me in the paddock at Brooklands during the September 1938 meeting and said that Mr Lyons had asked him to see me. Laurie Hathaway was leaving SS to work for Coventry Climax, and there was an interesting vacancy in the engineering department at Coventry. Would I be interested in joining the company to become Chief Experimental Engineer? Naturally this was something quite new, but after I had been up to Coventry to talk to Mr Lyons and Bill Heynes I decided to take the big step. For the first time since my marriage in 1933, I was ready to move our home away from London, and because SS already had a fine reputation in the industry the new job sounded very interesting. Little did I know then that Hathaway's movements were to benefit me again, in 1950.

I moved up to my new job at Foleshill, in Coventry, towards the end of 1938, and soon found that although the SS engineering resources were quite limited they still achieved remarkable things in a very short time. I had no responsibilities on the body engineering side – Lyons was the styling genius, while Harry Gill and Bill Thornton (ex-Triumph) were his principal body engineers. Bill Heynes was Chief Engineer, of course, though Harry Weslake had done the original conversion from side-valve to overhead-valve on the 2½-litre engine. All in all I doubt if there could have been more than half-a-dozen draughtsmen involved in mechanical design. I believe I was to be paid something like £12 a week, which would be well below starvation level now, but which made us quite comfortably off in 1938.

My first major project at SS was to develop a new independent front suspension designed by Bill Heynes. When I arrived all the parts had been made but were just lying in a heap on the workshop floor. I also had to improve Sammy Newsome's racing SS100 for use at Shelsley Walsh, and we also prepared a saloon for Jack Harrop and George Mangoletsi to drive in the 1939 Monte Carlo Rally. In fact the independent suspension was not fitted to a production car, even in modified form, until 1948 – on the XK120 and the

Mark V – because of the break in car production during the war.

In the spring of 1939 I moved the family into a rented house in Keresley Road, which was conveniently close to the factory, although for a time I had lived in digs, and perhaps it was this, combined with the fact that I was a cockney, which explained why I did not make many new friends at first. As I write this thirty-five years later it is difficult to believe that there was ever a time when I knew so few people in the area. I had met quite a few people at Brooklands who hailed from the Midlands, but apart from them I came to know few new people until I moved down to Bristol in 1940. I was always very busy in my job, and until Ethel and the children moved to Keresley Road I spent every weekend back in London.

Once established in Coventry, our closest friends were the Mundys, and here I must record the start of my long and still strong friendship with Harry Mundy, who seems to have been with me on so many happy and successful occasions in the last twenty years. Harry was at Alvis for some time, deeply involved in engine and gearbox design, before he moved to ERA to work with Peter Berthon and Murray Jamieson on engine design for the new Raymond Mays road car. I actually left ERA on the Friday to return to Brooklands, before Harry joined ERA on the Monday, so in Harry's words, "the first time we met we didn't!" However, we soon met through the Gray brothers, who did all the bodywork and panel-beating for ERA and were well-known at Brooklands, and we have been close friends and good colleagues ever since.

Most Coventry people, however, seemed to go round in tight social circles, and I discovered later that although there was some contact between people in different car-making companies, they maintained quite good security and rarely discussed the work they were doing. I suppose too much social chat could have let slip what they were planning, but in a trade where component suppliers' representatives moved freely from factory to factory it was inevitable that we should get to know what was brewing anyway! Motor cars in the 1930s were still changing from year to year, especially in regard to engine design, but they all seemed to follow their own standard design pattern, so there was rarely any need for what we should now call 'industrial espionage'. In any case, many firms took their cars to Brooklands for high-speed tests, and word soon got round of new developments.

When I was still working at Brooklands, for instance, Riley – who retained Reid Railton as a consultant – sent down a new $2\frac{1}{2}$-litre 'Big Four' saloon for our assessment. Riley wanted his opinion as to whether it would be good enough to do 100 miles in an hour, so Railton called me into his office, winked, and suggested that I should let him know. So of course I took it out, and I completed 100 miles in the hour on the track; that way Riley had an immediate answer!

At the time I suppose we all realised that war was in the air, though since none of us could do anything to delay it we tried not to think about it. Even when I was visiting factories in 1938 to collect parts for the Railton record car there was evidence of rearmament – at Beans' Industries, for instance, who were making hubs and wishbones for the car, it was possible to see thousands of shells lined up waiting to be delivered. Almost as soon as war broke out, all car production at SS stopped, and we began to make components for Armstrong-Siddeley, at the other side of Coventry, who were connected both with engine and airframe manufacture. SS were going to make aero-engine parts, wings, bomb doors and the like. I must admit that I wasn't at all interested in that sort of thing – there was no development work to be done, anyway, as Armstrong-Siddeley would look after that – so I busied myself motorising machine tools to help in the production drive. Previously they had been belt-driven from overhead shafting, but I converted machines to electric power with individual motors. There were a lot of new machines, because in that last year making cars we had survived with very little machinery at all. I remember that I was responsible for bringing the very first lathe into the experimental department, and there were only two old milling machines for making brake-rod fork ends.

I got on very well with Lyons and Heynes, but even so it didn't look as if there would be much interest for me at SS for the duration. Nor, it seemed, would there be much interest in other car companies in Coventry, so I started to look around for another more satisfying job. At the time no-one could know how long the war would last, so at first I was prepared to leave the family in Coventry, and move back when the war was over; there was absolutely no thought of losing!

Naturally I was keen to get involved in the war effort, so I looked round for development work connected with engines – by then I had become thoroughly immersed in engine work in preference to other more general chassis development – and after writing to various firms I was called down to Patchway, Bristol, for an interview with the Bristol Aeroplane Company. Bristol were expanding rapidly at the time, and I was offered the job of Technical Assistant on carburettor development for aero engines – something that was partly desk work, but very much in my line. As usual I was more interested in the technicalities of the job rather than any domestic difficulties it might cause, and I accepted it without too much hesitation. Yet again, therefore, for the third time in five years, I was moving into a job which would mean again moving house. We were lucky in that Charles Newcombe, an ex-Birkin mechanic, was at Bristol running the experimental machine shop, and he was able to find us a house at Downend.

Going to work for a large aero-engine firm was all rather strange at first. Apart from the sheer size, they seemed to be so very much more organised than any of the small companies I had known previously. However, even in wartime it took such a time to get things done, as there were so many systems and so many other people to be consulted before a modification could be made. As I had been used to taking many of my own decisions I found this rather frustrating, but eventually I began to appreciate the need for tidy administration in such a company. Much of my work was involved with detail carburettor development and the rectification of problems. Whatever we decided should be done, or no matter how tiny the story of a failure, we had to report on it. The report might be only one or two pages long, but the circulation list might involve hundreds of people.

It wasn't enough just to write down the facts of a case and have done with it, either. Bristol were very particular about the type of King's English they distributed, and for a while I used to have my reports thrown back at me for re-writing. Not only Bristol personnel but Air Force officers and AID people were involved, and it was extremely important that the story of failures, rectifications and modifications should be presented in such a way that everyone would understand thoroughly. At first I didn't like doing re-writes when I might have been working on engines instead, but eventually I came to see the wisdom of it all.

The carburettors were much more complicated than anything I had seen previously, even on a race track, and it was months before I absorbed all the details. It was at Bristol that I first met Phil Weaver, who was doing a similar job, but later made his own name as manager of the famous Jaguar racing department which produced the Le Mans-winning 'C' Types and 'D' Types in the 1950s. The engines were all supercharged radials, like the Pegasus and the later sleeve-valved series of Hercules, Taurus and Centaurus. The carburettors were made by Claudel-Hobson (who had a large factory in Wolverhampton, but had also taken over the old Triumph works in Coventry) and it was lucky for me that Phil knew them inside-out.

Much of our work fell into the category of rectifying problems in service, and there was a lot of travelling around involved. Bristol had their own test airfield nearby, as well as several repair factories in cigarette bonded warehouses, and the central electricity building in the City, all of which had been requisitioned for their use. Bristol itself was an important sea port, so it came in for its share of bombing on that score as well as because it was the home of the aeroplane company. One particular daylight raid was especially unpleasant as the bombers had to overcome very little defence and they laid a lot of explosive all over the airfield. We were pinned underground for more than three hours, which was hell, but the tragedy was that two or three shelters were hit, and one of the casualties was young Adrian Squire, who was working for Bristol by then. He had been the proprietor of that little firm on Remenham Hill, near Henley, which made a handful of really exquisite Anzani-engined sports cars in the 1930s.

After the bombing of the factory we had to walk home at about lunchtime, leaving our cars behind in the car park. There was so much confusion, fire and wreckage about that we couldn't reach them. Before long, however, Phil Weaver and I decided to go back and try to retrieve them. There were armed guards everywhere when we arrived, unexploded bombs sticking up out of the concrete, and craters all over the place, but we took our chance, got our cars out and drove home. It was well that we did. The following morning we arrived to find an enormous slab of foot-thick concrete where my car had been standing – the little Austin would have been flattened!

At the factory there was an early-warning system, when lights and a buzzer went off with variations to indicate how near the raiders were to the factory. We soon got to understand the code, but after that one disastrous raid we reckoned we might be safer a little further from Patchway than in the bomb shelters. We all had cycles then, and used to scorch up the road to a village about three miles away when the warnings went off. As these were often during pub-opening hours it used to necessitate some suitable refreshment while waiting for the all-clear to sound. We called ourselves the Patchway Harriers!

Though I can't say I was very happy at Bristol at first, it certainly wasn't because I knew no-one. People like Harry Mundy, who had joined the RAF, always managed to find time to call in for a talk when duty brought them to the company, and among the other people who worked there were H.G. Tyrrell-Smith, the Triumph motorcycle man, Mr Bush, who was in charge of our office and had been at Daimler in the 1930s, and Francis Beart, who made a great name for himself as a tuner of Norton racing engines.

In spite of all the bombing and the general miseries of the time, Ethel and I agreed that she would be better off near me than in Coventry, and in the worst possible national atmosphere we took the house in Downend. It was our sixth home since we were married in 1933, but we enjoyed our stay in Bristol very much. Moving the family was more than usually strenuous because Ethel was expecting the arrival of our third child – Richard – who was born in Bristol soon afterwards; John was now six and Peter was four years old.

However, I was determined to get back into the car-making business as soon as possible, and I made sure I kept in close touch with Bill Heynes. One day in 1942 I heard from him with the news that SS Cars had been in contact with the Ministry of Supply who required a variety of small vehicles to be developed specially for parachute dropping. It sounded much more like my sort of engineering than the work at Bristol, and I was thrilled to know that I was wanted back in Coventry. But the fact that I wanted to go and SS Cars wanted me to go was one thing; actually winning a battle for release from my existing job was another. Under the many wartime regulations there was complete direction of labour, which meant that one simply couldn't change one's job just like that.

I suppose I must have been of value to Bristol because they certainly didn't want me to leave. First I had to get letters from SS to support my request for the move and to explain the work they were doing, then I had to see my immediate boss, Roger Ninnes, and finally the Technical Director of Bristol, Roy Fedden. They all listened to my tale of woe, and eventually I was allowed to leave. That should have been the end of the Bristol story. However, a couple of years later, Roy Fedden approached me obliquely and asked me to travel down to Cheltenham to look at a new car he was planning to produce after the war. Fedden had already dabbled with car production many years earlier (this was the Phoenix), but this project was to be all-new, called the Fedden, and would have a radial engine mounted at the rear. There was no car to look at when I visited him, but a rather beautiful brochure which showed how the car would be built. There was also a Czech Tatra, which was much the same sort of car, and surprisingly advanced when first announced, with virtually the same layout apart from an air-cooled V-8 engine. It was all very interesting, and technically exciting, but I turned it down because somehow I didn't think it would succeed. In that I was quite right, though Fedden put a lot of effort and money into it from 1945 onwards. The radial engine was a big mistake, I felt, and the weight distribution was never conducive to good handling. Even though some of the finest engineers in the country worked for Fedden on the car - Gordon Wilkins styled it, Ian Duncan was Fedden's assistant who later founded the Duncan body-making concern, and Peter Ware was later Technical Director of Rootes when they produced the Hillman Imp - it was never put into production.

So I left Bristol after a short flirtation with the aircraft industry. I was looking forward to getting back into the swing of things at SS, even if the vehicles involved would not be of a sporting character. In 1942 I could have no idea of what William Lyons had in mind for his post-war cars. Lyons and Heynes, however, had much more on their minds than the fighting vehicles and aircraft fuselages which were being made in ever-increasing numbers at Foleshill. They had nothing less than a complete new range of products in view, and I soon found out that I was to be involved from the very beginning. Fire-watching was to take on a new slant - for it was on those otherwise-boring nocturnal occasions that the post-war Jaguars took shape.

Napier-Railton Land Speed Record car

Reid Railton's beautifully-streamlined twin-engined masterpiece was most certainly the most advanced land vehicle of all time when it was unveiled in the spring of 1938. To use a Pomeroy-like phrase, its design was 'elegant' as well as purposeful, and not even the latest in Mercedes or Auto-Union Grand Prix cars was as nicely engineered. By comparison it made all previous LSR contenders look like streamlined lorries.

Railton's philosophy was to start from a near-perfect tear-drop shape, and pack the necessary power, fuel, systems and a driver inside it. Although it must not be confused with John Cobb's other record car, the 'conventional' Napier-Railton which he used at Brooklands and elsewhere for record-breaking endurance runs, this Railton also used Napier Lion engine power, but in this case there were two engines and they were supercharged.

The mechanical layout was quite startling. The main frame was a big, tubular-section, S-shaped backbone bounded more or less by front and rear wheel centres. The intrepid John Cobb sat ahead of the front wheels, and behind the rear wheels there was only fresh air and some support for the streamlined tail panels. The backbone was S-shaped so that the two W-12 Napier engines could be arranged in echelon. The forward engine drove the back wheels, the rear engine drove the front wheels, and there was no interconnection between them; this must have been one of the very first vehicles to have completely independent, non-connected, all-wheel drive. There were two gearboxes, two freewheels, but no clutches. The body was completely enveloping, and to effect wheel changes between runs it had to be removed completely as there were no wheel-access panels.

Several companies had a major involvement in the construction of the Railton, and although Thomson and Taylor were responsible for assembly, John Thompson Pressings made the main chassis frame, David Brown Industries the transmissions and Lockheed the brakes. Hassan had to deal with these and many other suppliers, including Smiths (instruments), Lucas (electrics), Dunlop (tyres and wheels) and Burman (steering gear).

The car was sent to Utah in the summer of 1938, where it was soon locked in battle, from testing and practice to the actual runs themselves, with George Eyston's 5,000 hp Thunderbolt. It is worth recalling that the Railton's power output was approximately 2,500 bhp, its weight 62.5 cwt, and its engine capacity 47,872 cc, while Thunderbolt produced about 5,000 bhp, weighed approximately 140 cwt, and had an engine capacity of 73,164 cc! After a considerable wait for ideal salt and weather conditions, Eyston's car set the pace with a two-way run of 345.5 mph on August 27th. Cobb countered with 350.2 mph on September 15th, and the very next day Eyston went out again and lifted the record to 357.5 mph.

Both cars returned home for modifications aimed at raising the speed again in 1939, but Eyston's car never re-appeared at the salt flats following its big weight-reduction programme. Instead it was shipped out to an exhibition in New Zealand in the autumn of 1939, left there for the duration of the war, and tragically was destroyed by fire in 1946 at Wellington airport, when a warehouse in which the unique car was stored was burnt out.

The Railton was improved in detail for 1939, and in the summer, with war clouds already looming over Europe, the same expedition arrived at Utah. With the minimum of fuss, once good conditions had been achieved, Cobb raised the record to 369.74 mph on August 23rd, and three days later raised the 5-km, 10-km and 10-mile records as well. Three hours after the last record was in the bag there was a rainstorm and thē salt surface was turned into a lake! A week later war was declared, and since there was no point in risking the car's return passage by ship it was taken to Canada for storage.

Eight years later, the old car – old only in years but not in usage – was resurrected, re-worked in various details by Reid Railton, who by then had emigrated to the West Coast of America, and complete with a new sponsor – Mobil – was sent again to Utah. On September 16th, 1947 the car, then known as the Railton

Mobil Special, set up a record of 394.2 mph for the two-way run (and a best one-way run of 403 mph) which was to stand for a further seventeen years. Only a multi-million pound assault, and several years of frustrations, allowed Donald Campbell's Bluebird-Proteus to raise this by a further nine mph.

Following the 1947 record the car was shown by its sponsors in America, then returned home to Britain, eventually to find its way to the Birmingham Museum of Science and Technology, where it still rests.

4
Birth of the XK120

Although I had learned a great deal in my short stay in Bristol, it was a great pleasure to return to the motor industry again. Back at SS, in Foleshill, I was soon thoroughly absorbed in the fascinating war work which SS had collected in my absence. In the assembly shops where SS-Jaguar cars had been made in 1938 and 1939 there were ever-increasing numbers of aircraft wings, main fuselage sections and other components.

My work, however, was to be only indirectly concerned with aircraft. The Ministry of Supply were determined to develop a range of lightweight vehicles which could be dropped by parachute alongside airborne troops, land without damage, and be driven off straight away without further assembly. SS had experimental contracts with the Government to develop small vehicles, and our little four-man department was set to work on them. We built some rather odd vehicles, among which were lightweight trailers to be towed behind motorcycles and jeeps, mule carts for the Burma campaign, a folding sidecar, and later still a couple of small Jeep-like vehicles for air-dropping.

We modified a handcart so that it could be towed by mules, and one day we were very surprised to receive a visit from three chaps straight out of Burma; they were all complete characters (in that campaign you just had to be) and one of them was dressed in the kilt and carried a shepherd's crook kind of thing which was taller than himself. They were involved in getting things organised for the next big push forward, and knew exactly what they wanted. They knew precisely what the terrain was like and how wide the jungle paths were – something we could only guess at from England. We had to modify the cart considerably for them, but in the end they were satisfied. It would have been nice to carry out tests behind a mule, but that was something we didn't have at Foleshill in 1943, and I couldn't persuade anyone to buy one for the company to use!

The project which really interested us was the lightweight car. As well as being very light and not top-heavy like a Jeep, it had to be very simply engineered, capable of being driven over very rough terrain, and of course strong enough to withstand parachute drops on to rough ground. The first car or

'Jeep' we made was a very simple thing, with most of the weight concentrated over the back wheels to get traction, because we could only consider two-wheel drive. It was fitted with an air-cooled twin-cylinder engine and performed reasonably well, but it wasn't considered powerful enough, so the second prototype was built with a side-valve Ford Ten engine in rather more orthodox form.

Unfortunately for us, by the time the second car with the Ford engine was performing well, the Royal Air Force had learned how to drop full-sized Jeeps by parachute without damaging them, their aircraft were getting big enough to transport them, and of course the Jeep had a fantastic reputation for reliability. The Army let it be known that they would rather use a smaller number of Jeeps than a large number of lightweight imitations. From time to time we had to take the little vehicles down to the Wheeled Vehicle Establishment at Farnborough for comparative tests against other designs – Standard had their little 'Bug' which appeared again in the 1960s in the London Motor Club's 'autopoint' TV races, Morris (with Sir Miles Thomas in attendance) had a similar project, and someone else had a modified motorcycle and sidecar. We never actually produced any more, but the Establishment bought our vehicle for assessment. They must have used it from time to time because many years later they actually wrote to Jaguar and asked for some spare parts!

There was one hilarious occasion when we were asked to attend air-dropping tests at an airfield in Yorkshire. Apart from the drops themselves we were to look at the bases upon which the vehicles were to be mounted for the drops. These were made by Triang, the toy makers, but the intention was that the winning contractor for the vehicles would also make his own bases. These were built as shock absorbers, and would have strips of metal running round rollers so that when the thing tended to collapse the distortion of metal around the rollers would be converted into heat and dissipate the energy.

The first drop was of a full-sized Jeep under seven or eight parachutes, which went off well, as the landing was fairly soft, but this was followed by a drop under only three 'chutes which was most spectacular because the Jeep landed very heavily having already virtually broken its back when the parachutes opened. Certainly it was in no fit state to be driven after the drop, and it took three drops in all before a Jeep was left in a drivable state. The drops for our own precious prototypes were cancelled, and I don't think the SS lightweight vehicle was ever subjected to the treatment for which it was designed.

Once the heat had gone out of the experimental vehicle development I spent more of my time looking after the tooling which we made to assist manufacture of the aircraft wings, frames, bomb doors and the like which were pouring out of the factory. Many of the tools were designed to help unskilled labour – or rather I should say new labour, many of whom were women – and because a lot of the aircraft panels were made in relatively small numbers this involved hand-beating over formers of complex section, which often needed a lot of minor modification and refinement before they could produce a satisfactory end result.

The job was interesting enough, but my enthusiasm by now was completely captured by what was going on at the factory out of working hours! For all of us, a spell of fire-watching was compulsory every week - which meant that we had to spend one night out of bed as unpaid wardens keeping a look-out for incendiary bombs which might drop on the SS factory or on surrounding houses, then raise the alarm and try to do something to minimise the danger. Fortunately, air raids on Coventry had just about stopped by 1943, so without something interesting to occupy us, these fire-watching sessions would have been a tedious and tiring bore.

But William Lyons, never a man to miss an opportunity, decided that this was a heaven-sent opportunity to get a bit of planning and discussion going when we were not tremendously busy on essential war work. He made sure that the fire-watching team on Sunday nights would comprise Bill Heynes, Claude Baily, myself and himself, and he turned these sessions into a sort of running 'design seminar' for what SS would do when car production could be resumed after the war. We would foregather at the works at about 6 pm, sign on, and while keeping an eye on the job, as it were, Mr Lyons would then settle down and talk to us about the future. During my first sojourn with SS, I had always found Mr Lyons to be a rather austere sort of man, very formal in his business contacts, although very interested and very interesting on the question of car design. He had not been much involved with me personally then, though we used to see him in the experimental departments every week. During those Sunday nights, however, he unbent quite a lot, and I think he got to know us much better. Even so I never heard him call anyone by their Christian name - no-one, that is, except Fred Gardner, who ran the sawmill and built all the prototype body shapes for him. Mr Heynes was always 'Heynes', and Mr Whitaker always 'Whitaker', even after they had been with him for many years.

But that doesn't mean that William Lyons was unthinking about his staff - far from it. During the war, when things were very tight, he sometimes threw a garden party in the grounds of his house at Wappenbury, where he always put on a good spread and made the children very welcome. On one occasion the Hassan family - Ethel and the two boys and myself - had cycled over from Whitnash, but Mr Lyons lent us his car to take home and the bikes had to go in the boot!

He was not trained as an engineer, but he had a very clear idea of the kind of car he wanted to sell, and was always very critical of deficiencies in performance, roadholding, suspension, noise and comfort. But while he was never slow to tell us - in no uncertain terms - when he didn't like something, he would leave it to us to sort out the solution. During the blitz he had been working away on body designs - he was always the stylist as well as the company boss - and along with Fred Gardner and a chap called Holland they found time in odd moments and evenings to make up several 'half bodies' - split down the centre-line of the shell – of various shapes. These developed in a logical way from the SS-Jaguars made in 1939, and really the XK-engined Mark VII saloon was the long-term result of this styling work, although it didn't appear until 1950, and the 1948 Mark V was the car most obviously descended from the pre-war models.

During these fire-watching sessions we also discussed engine requirements. The engines used in SS-Jaguars up until then were all basically Standards, with overhead-valve conversions, made in three sizes of 1,776 cc (four-cylinder), 2,664 cc and 3,485 cc (six-cylinder), but Mr Lyons wanted to cover this range with a couple of new engines of his own, which were to be more advanced, more refined and rather more powerful. The interesting point is that we confined our discussions to the bigger 'six' at first, yet the first engines built were 'fours'; at the time we were quite sure that fuel consumption and vehicle taxation would be terribly critical after the war. The cylinder-bore dimension was still the determining factor in the 'horsepower' rating, and for quite some time we worked on the idea of a new $2\frac{1}{2}$-litre engine to replace both of the Standard-built 'sixes'. It was thought that $2\frac{1}{2}$-litres would be as big as people could afford because they would be so short of money and fuel after the war. So Mr Lyons asked us to think in that range, but I remember that he was insistent on having an engine which would look smart and powerful – "something like the old Peugeots and Sunbeams which raced".

He wanted an output of around 120 bhp and he had a great liking for a twin-overhead-camshaft layout, but Claude Baily and I pointed out that this would be expensive and probably fairly noisy, too. On the evidence of the engines we used pre-war, we felt we could design an engine with a simple bathtub cylinder-head combustion space and with push-rod overhead valves, which would give him all the power he wanted. As far as we could see this would be easier, cheaper to make, and quieter. However, this didn't satisfy him at all, and if he wasn't completely satisfied with anything he would never agree to it. His new engine would have to be good-looking, with all the glamour of the famous engines produced for racing in previous years, so that when you opened the bonnet of a post-war Jaguar you would be looking at power and be impressed. He got his way, of course – Mr Lyons always did! – but I must admit that the rest of us thought it was rather a waste of time and money at that time.

So that is how the post-war engine studies took place. In a little development department office at Foleshill, with camp beds rigged up for our use, the first lines were drawn on paper for prototypes of the XK engines which are still in successful full-scale production thirty years later. Bill Heynes, as Chief Engineer and Technical Director, was in overall command; Bob Knight (another future Technical Director at Jaguar) busied himself with chassis refinements, including the developed version of the torsion-bar independent front suspension I had built in 1938; our Chief Engine Designer was Claude Baily, who had moved over from the Morris Engines factory; and I was to be in charge of engine development once we could spend some time on the test beds.

The first design to be made up, which took a long time because of the difficulty in getting non-war-effort engineering done at the time, was coded the XF (XA to XE were all experimental projects but were not actually built) and was a purpose-built four-cylinder engine of only 1,360 cc. I really don't know why it was so small as it had been schemed before I returned from Bristol (the six-cylinder version would have been only 2,040 cc, so quite a bit too small for Mr Lyons' plans), but in many ways it was a true forerunner of the XK engine. It had a stroke of 98 mm, which persisted for some years on other experimental engines until we agreed on the need for more low-speed torque, achieved by increasing the stroke to 106 mm. It was intended to prove out our ideas on twin-cam cylinder-head design, which used as simple a form of 'Ballot-type' valve operation as we could devise, and succeeded admirably. The bare bones of this design were settled even before I returned from Bristol, although we didn't get an engine running until later in the war. It gave us a lot of the answers we needed to incorporate into the final designs, and the basic cylinder-head layout was to be carried over virtually unchanged, except in dimensions.

However, just after the end of the war, in 1945, we decided to make another prototype – the XG – which could possibly be much cheaper to build, and would certainly be much less costly to tool up for production. The idea was that we should still design a proper hemispherical head with inclined valves, but that the whole thing should have a 'BMW 328' layout with the valves operated from a single side camshaft in the cylinder block. The engine would look as smart as Mr Lyons wanted, but it would have vertical carburettors sticking up above the cylinder-head as there was to be a vertical inlet port. The inlet valves would have more-or-less direct actuation from the side camshaft, while the exhaust valves would have extra pushrods across the cylinder-head and extra rockers in BMW style. Before the war I doubt if we would have been able to use this design because of patent restrictions, but as Frazer-Nash and Bristol took over the design in the aftermath of the fighting it became 'fair game' for development.

For the basis of the first engine we used the existing SS four-cylinder 1,776 cc cylinder block, crankshaft and moving parts, which were readily available and familiar to SS-Jaguar users, even though they were not advanced by our post-war standards. I never liked this engine on engineering grounds or on looks, and was happy that our experience with the single prototype proved that it was difficult to keep the valve gear quiet, and that air flow through the vertical inlet ports was not as high as could be achieved through the normal horizontal ports in our other designs. That last fact wasn't enough to deter Mercedes-Benz from adopting the same principle for its 1954/55 Grand Prix straight-eight (an engine which was to influence my later work considerably!), nor did it deter Bill Heynes and Claude Baily from trying the same layout again on their 12-cylinder designs in the early 1960s. We have, however, proved conclusively by air flow experiments that this form of inlet port is not as efficient as the conventional side-flow port for producing maximum power.

Eventually, Mr Lyons decided that we should design our new engine range from scratch without being hampered by existing tooling or using carry-over parts. In the meantime he went along to see Standard's boss, Sir John Black, and bought all the six-cylinder tooling and arranged to ship it up to Foleshill as soon as possible. He also arranged for further supplies of the so-called 1½-litre SS engine from Standard – at 1,776 cc it was not really a 1½-litre at all – but he couldn't buy the tooling as Sir John Black wanted engines to put into his new Triumph 1800 roadster and saloon models. That was in 1945, and it meant that the first post-war Jaguars could use existing pre-war engines while we all got on with the development of the new engine.

The next prototype was really the final answer, though we had to put in a lot of work before it was ready for production at the beginning of 1949. We started scheming and detailing during 1945 and the engines were certainly running in 1946. Mr Lyons had come round to the view that we needed more than just one engine to replace all three old ones, and after a lot of discussion we decided to go for the very logical 'family' of a four-cylinder and a six-cylinder design. The new tooling conceivably could make both types of engine, and many of the parts and processes – valve gear, cylinder-head machining, block machining, pistons and connecting rods – might be common. The engines were coded XJs, and at first we stuck to the 98 mm stroke of the first XF engine. Now that we were to have two engines we could abandon the 2½-litre concept to cover the whole range, and by using different cylinder bores we made the 'four' into a 2-litre (1,996 cc actually) and the 'six' into 3,182 cc. The development work that went into these designs was carried out on both versions, and there was a variety of experimentation on valve gear and camshaft details. One of our biggest problems at first was to get the chain drive to the camshafts quiet enough, and another was in stopping oil leaks around the camshafts; at the time the timing case was quite separate from the cylinder-head, so the camshaft poked through from one to the other in the wind, which gave oil a splendid chance to leak if it felt like it. It usually did!

All this time I not only had to keep development of the engines in hand but also had to oversee the chassis work, too. I was in charge of development of all mechanical items, but in those days of separate coachbuilt body-shells I had nothing at all to do with the bodies. Because of all the money Mr Lyons was pouring into the new engine, he decided that he couldn't afford a completely new car as well. Therefore the Mark V, which arrived in 1948, still had a separate body-shell built up and welded at Foleshill from lots of small pressings on rather simple jigs, though beneath it was a chassis which finally incorporated the torsion-bar front suspension which had been occupying our attention for ten years! We made umpteen prototypes with slightly different details, and were very pleased when the final torsion-bar set-up was agreed. All the work certainly wasn't in vain for the same basic suspension was used on XKs until 1961, and the same basic system was then adopted for the 'E' Types.

We did a tremendous amount of development on the new twin-cam engine, which became the XK with all the refinements, the longer stroke, and the trimming-off of corners and smoothing-out of exterior contours to make sure it looked really smart. For a long time Mr Lyons was serious about productionising the four-cylinder version of this engine but however hard we tried, when we installed it in a car there was always evidence of the secondary vibrations which one simply can't eliminate from a big 'four', which manifested itself as a 'zizz' through the transmission and structure generally. If it had been possible to sell the XK two-seater at the lower price which it would have to have carried with the smaller engine, and still make a profit, then I suppose we would have struggled on with development, but the smaller Jaguars never made much money for the company, and I understand that Mr Lyons was not sorry to concentrate on larger-engined cars from 1948 onwards!

In terms of power output and general efficiency the 2-litre XJ engine was very satisfactory. We race-tuned one so that 'Goldie' Gardner could install it in the Gardner Special record car, which was originally an MG, EX135. The tuned 2-litre produced 146 bhp for the Special, and was tall enough to call for a hump in the front panel of the streamlined shell. It propelled Gardner's little car to a record-breaking 176.69 mph on the Jabbeke road in Belgium, and I remember that MG enthusiasts who had always thought of this car as an MG were a bit shocked, though they were quick to point out that the same car had achieved 203 mph with an 1,100 cc engine in 1939. What they completely failed to point out at the same time was that the MG was heavily supercharged, whereas 'our' engine was not!

When the XK six-cylinder engine was released to the public, both it and the beautiful XK120 car it powered came as a complete surprise. Both had been a very well-kept secret, which was surprising as there were the same sort of problems with travelling representatives from supplier companies as there had always been; however, by this time they seemed to keep faith rather better than in the 1930s and security was well preserved. One must also remember that the XK engine was not a worldwide first - my old boss W.O. Bentley had designed a very nice 2.6-litre twin-cam for Lagonda to use, and there was also a Meadows-built twin-cam for the Invicta Black Prince. Both of them had been announced prior to our own, though the Invicta design was never a success, and the Lagonda had to wait for the company's takeover by Aston Martin before it became famous.

Mr Lyons really needed his engine for large-scale production in a new saloon car, the Mark VII, but this wouldn't be ready until 1950. He certainly didn't want to sell it in the 'old-fashioned' Mark V, though he did build a couple of cars in this form, which we always called the Mark VI, hence the reason why there never was a production Mark VI Jaguar! So the engine was to be seen first in a sports car, a sports car, moreover, that was never intended to sell in large numbers. Of course, the XK120 - the name referred to the XK engine and a top speed of 120 mph - was an overwhelming success, and the

What was that about lightweight fighting vehicles? I demonstrate the manoeuvrability of Jaguar's VA prototype

The Ford-engined Jaguar prototype looked more like a Jeep than the VA, but it also failed to earn a production contract

Checking the 2-litre prototype four-cylinder XK engine installed in Goldie Gardner's Special, which achieved 176.69 mph on the Jabbeke highway in Belgium in 1948. Gardner is in the linen helmet

Red, white and blue Jaguar XK120s lined up at Silverstone in 1949, when they dominated the Production Car race. Leslie Johnson's white car (HKV 500) won at 82.2 mph and Peter Walker's red car finished just a few seconds behind. I was much slimmer in those days

story of how we were obliged to convert the two-seater body for large-quantity production has been told many times. The plan to offer an XK100 version with the four-cylinder engine was shelved when development work on the 2-litre engine ended.

Ron Sutton ('Soapy' Sutton to his friends – and I won't say how the nickname came about!) took a nicely prepared XK120 to Belgium in May 1949 and achieved 132.6 mph, which was outstanding for such a refined sports car. Maximum refinement had been our goal from the start, even though the XK120 was probably the most efficient engine in Britain in terms of power output per litre. This involved a lot of work on intake silencing, the chain drive to the cams and the exhaust systems. Claude Baily and I always had a very good idea of the race-tuning capability which was being built into the XK engine during its design, but even though it looked like a racing engine it was never intended to be one, and it was really quite by chance that Mr Lyons seized on the opportunity to prove how good his new sports cars were by entering a team in the Production Car Race at Silverstone in 1949. He was persuaded that here was a marvellous opportunity to prove the cars' supremacy, but first he had to have it proved to himself!

We had to hire Silverstone for a secret session, to show him not only that the car was quick enough but that it would also last the distance (it was only for one hour's duration but he wanted to be sure). Mr Lyons, Bill Heynes, 'Lofty' England, Bill Rankin and I went over to the circuit one summer afternoon and set to work. 'Lofty' and I had to do all the driving, and although neither of us was a racing driver we had to break the circuit record for that class of car to satisfy Mr Lyons. 'Lofty' had a go, then I had a go, and we still hadn't quite reached the lap speed we wanted, so I kept going round and round until I overdid it and went through the straw bales! But I kept going, and in the end I reckon I went through the bales at least three times by which time the front end of the car – it was still a bit precious and just about the only one we had – was getting well and truly battered and looking like a rather rapid haystack.

Finally we achieved the speed and the mileage, and we hoped Mr Lyons would be satisfied. Then suddenly he said that he would like to try it himself, and commanded poor Bill Rankin (Jaguar's Publicity Manager for many years, and one of the founder members of the SS Car Club) to get in and go round with him. Now Mr Lyons was no slouch in a car – he and Rankin had shared a car in pre-war rallies – but over the years his faculties had changed just a little. No sooner were they in the car and off down the track towards the first corner than Mr Lyons said: "Rankin, tell me when to brake, I have left my spectacles at home" – he was effectively driving blind! Rankin had to thump him in the back and tell him to start turning to left or right and they did about two laps like this, but by this time Mr Lyons was satisfied and pulled in looking happy, while Rankin looked like a piece of chewed string!

It was all worthwhile in the end, because we appeared at Silverstone with three XK120s – one each patriotically red, white and blue – and Leslie

Johnson's white one won the event with 82.2 miles covered in the hour with Peter Walker's red car just a few seconds behind. Bira drove the blue one, and led for a time, but eventually thumped the straw bales following a puncture and bent it. Norman Culpan's 2-litre Frazer-Nash hung on magnificently to finish third. The winner was the same car that had achieved 132.6 mph in Belgium when, incidentally, I would have been driving it had I not been laid low with mumps at the time. It was a very successful car, if not ultimately as famous as Ian Appleyard's rally XK120, and apart from winning the Silverstone race Tommy Wisdom and I took it for a recce of the 1950 Alpine Rally area to assess the car's chances for Ian Appleyard, and later still Johnny Claes used it to win the Liège-Rome-Liège rally in 1951.

At the end of 1949, I had time to look around and decide how I was getting on in my career. When the family returned from Downend to Coventry in 1943 we bought a house in Whitnash, a suburb of Leamington, and settled down very happily. However, there was no doubt that I was also very settled in my job at Jaguar, so settled in fact that it didn't seem as if I could look forward to promotion there for some time. I was already Chief Development Engineer on the mechanical side, but alongside me as Chief Engine Designer was Claude Baily, who was no older than me, and our Chief Engineer was Bill Heynes, who was just a couple of years older. We had a very satisfactory working relationship, but it was obvious to all that the status quo would exist for a good many years.

It was therefore a very pleasant surprise – one of several I have had in my working life – when early in 1950 I was approached by Laurie Hathaway (by then a consultant to Jaguar) who asked discreetly how happy I was? As already mentioned, I had moved into Hathaway's chair at SS in 1938 when he had moved on to Coventry Climax, and in recent years he had left Coventry Climax to set up in business on his own. Hathaway had originally been at Humber with Bill Heynes and was retained by Jaguar as a consultant almost as soon as he became self-employed. When I replied guardedly to Laurie's inquiries, he said that Mr Leonard Lee, who *was* Coventry Climax Engines, was looking for a new Chief Engineer, and would I be interested in going to see him about the job!

Of course this was wonderfully interesting news, even though I knew that there would be work other than on engines to be done, mainly in connection with fork-lift truck design and development, as for the first time I might be at the top of a team of engineers with no superiors apart from Mr Lee, who was Chairman, Managing Director *and* the major shareholder in his own private company.

I was a bit puzzled at first about the inquiry, because I knew that Coventry Climax had been managing for some time without a Chief Engineer at all, but this could hardly have been satisfactory. Laurie Hathaway had left the company some considerable time before this offer came along, and after his departure Mr Lee took on a new man in connection with gas turbine work which was then starting up at Widdrington Road. The Ministry of Supply had been anxious to find a firm willing to work on gas turbines smaller than those already used in aeroplanes, to gauge the scale effect and the cost of the thing. Coventry Climax had hired a man from Whittle's company (Power Jets, in Northampton) to be Chief Engineer, and the turbine work had come over with him. The turbines had been made and tested, but no production contracts had been placed, so that was the end of it.

However, it became clear that Climax's main interests would still be in fork-lift trucks and small piston engines, and when the gas turbine work was seen to be unproductive their Chief Engineer had moved on. Although Leonard Lee tried to carry on without a replacement for a time it proved increasingly difficult, and I went to see him – I had met him before – and was very pleased to be given the opportunity of joining him. Apart from any other considerations of power and influence, he offered me a salary of £2,500 a year, which for 1950 was real money indeed, and just about double that which I had been earning very happily at Jaguar.

Of course this was absolutely splendid news, but as I am normally a very loyal and open person I felt that I had to go and discuss this with Bill Heynes at Jaguar. I have never been a particularly ambitious or 'pushing' person, but on the other hand I have never neglected an opportunity. I spoke both to Heynes and Mr Lyons, and when I put it to them that I had been offered a very good new job on the other side of town at such an attractive salary they both agreed that I couldn't afford to ignore it. Of course, there was simply no way in which they could increase my salary to match the offer without upsetting a lot of other people in the company. They wished me well, if I should decide to go, and there was absolutely no ill-feeling at all.

It was typical of the relationship I had built up with Mr Lyons. He was always absolutely fair with his staff, though he always made sure that he got full value from them. He had a reputation for being a hard man in some quarters, but this was only because he demanded loyalty and as much effort from his managers as he was prepared to put in himself. There was never any waste of money at Jaguar and certainly never any surplus staff, but it was all very businesslike and I think that everyone who worked for him respected him for it.

By 1950 I had been with SS and Jaguar for more than seven years, and a move was going to be hard to make, but the attractions were so obvious that I didn't hesitate for long. In February 1950 I left Jaguar for what I was sure would be the last time (how wrong I was!) and started work at Widdrington Road. A world bounded by fine engines and fast cars was to be replaced by one concentrating on industrial engines and fork-lift trucks. It seemed at the time that I couldn't have been further away from motor racing but if only I had known what it would lead to in the next ten years!

Jaguar XF to XK engine development

We have already seen how the bare bones of post-war Jaguar engine design policy was thrashed out between Bill Heynes, Walter Hassan and Claude Baily, guided and encouraged by William Lyons, during those legendary fire-watching sessions at the SS Cars factories towards the end of the Second World War. In his job as engineer in charge of engine development, Hassan was well-placed to guide the team through the several versions of prototype tried before the final XK layout evolved.

As ever in considering the way in which very important changes took place in the British motor industry, it is necessary to trace the business and technical events which led to the decision-making. In 1939, when passenger car production ceased, SS Cars used three engines – the overhead-valve, four-cylinder 1,776 cc unit producing 65 bhp, the six-cylinder, 2,664 cc version of this, producing 102 bhp, and the more prestigious Heynes-developed 3,485 cc, 125 bhp example of the six-cylinder unit. All outputs are quoted 'gross', for the industry had not got around to quoting the power actually available for driving the car – it didn't sound impressive enough! The 1,776 cc and 2,664 cc engines were used in side-valve form for the Standard Flying 14 and the Flying 20 and were produced by the Standard Motor Company in modified form for SS Cars on special tooling at Standard's Canley factory.

Technically there was need for change (Lyons had originally used side-valve engines from Standard for his first SS1 coupé in 1931, and they hadn't advanced much since then apart from the overhead-valve conversion), but there was something much more personal involved. In 1931 Lyons was happy to let a much bigger firm – Standard – build chassis and engines on his behalf, but by 1939 his business had expanded so much that it seemed inappropriate to use sub-contractors, perhaps even competitors, in any volume. Further, both he and John Black of Standard had very strong business personalities, and it is generally known that there was little more than civil tolerance in their relationship by then. Indeed, Sir John Black wanted to 'get at' Lyons by building sports cars of his own after the war (hence his purchase of the bankrupt Triumph company), and for his part Lyons wanted to build his own engines. When, at the end of the war, Lyons made an offer to buy the special tooling for the six-cylinder engines, which was of no further use to Sir John Black, and the offer was accepted, Coventry sources say that Lyons immediately sent a messenger round to Canley with the agreed cheque in case Sir John should change his notoriously volatile mind!

When Bill Heynes read his famous paper ('The Jaguar Engine') to the Institute of Mechanical Engineers, in 1953, the political realities of the situation were obscured by good mechanical reasoning. Extracts from the opening paragraphs make clear the reasoning behind the eventual development of twin-cam units:

"During the years immediately after the 1939-45 war, it became increasingly obvious to the author's company that the power developed by its then current engine was being approached and, in certain cases, equalled by some of its competitors.... The existing range of engines had been developed as far as economically possible, and efforts to extract high horsepower and greater speed range brought out the inherent limiting factors which existed in this type of unit.... The market had been developed over the past years in two sizes of vehicle, and in both markets there had been success in meeting competition and establishing a strong following of enthusiastic owners, all in the medium price group. It was therefore obvious that the new range of engines must fall approximately in the same two groups in which the firm had built its reputation.... To enable tooling costs to be kept to a minimum it was desirable that as far as possible parts on the two units should be interchangeable.... This eventually led to a decision to go ahead with a four- and six-cylinder as the best proposition; first, because on these two engines common tooling can be accommodated and common parts employed, and secondly, because this followed directly in the market which had already been developed.... Among other items regarded as essential was the need for the engine to be capable of propelling a full-size saloon at a genuine 100 mph in standard form, and without special tuning.... There was just one more requirement which had to be met - one which automobile designers (on both sides of the Atlantic) as a general rule ignore - and that is the styling of the external design of the engine so that it looks the high-efficiency unit that it is...". Little more comment is needed - these words summed up precisely the aims and hoped-for achievements that Lyons demanded from his new engine.

The first experimental engine - the four-cylinder 1,360 cc XF design with its bore and stroke of 66.5 mm × 98 mm - was produced with the object of assessing the merits of the twin-overhead-camshaft design and cross-flow breathing which the fire-watching sessions had thrown up for development. Neither the cylinder block strength nor the crankshaft durability was found to be adequate, and the Heynes/Hassan team passed quickly on to the next experiments with the XG design, which was a further conversion of the old 1,776 cc Standard (73 mm × 106 mm), with what was to all intents and purposes a BMW 328 cross-pushrod type of cylinder-head. It retained the same side-camshaft position as before and the original cylinder block. There was a vertical inlet port which meant that carburettors would have had to be mounted overhead, and the effects on future Jaguar styling could be imagined! This design was dropped as being neither sufficiently powerful nor refined enough for Lyons' requirements.

The third step in the engine-development programme produced the XJ engines (not to be confused with the much more recent XJ12 engines, or indeed the XJ cars) in four-cylinder and six-cylinder forms. They shared a common stroke of 98 mm, and cylinder bores of 80.5 mm and 83 mm, respectively. Most of the serious development which Hassan undertook for the final engines took place on the four-cylinder engines, the six-cylinder units not being built until the basic design had been proved and finalised. When Hassan and Heynes decided that there was a need for higher torque at lower engine speeds (this was a theme that recurred many times in Hassan's working life), the simple way to achieve this was by a longer stroke. This could be accommodated in the existing design of crankcase, and for reasons not unconnected with the existing tooling which had just been purchased from Standard the new figure of 106 mm was chosen, but this was only to be applied to the six-cylinder engine as it was thought important that the small unit should stay within the 2-litre limit.

As we know, the four-cylinder engine never went into production, although it had the first 'public' showing, and the 3.4-litre XK engine was put into a sports car more-or-less as a publicity exercise, but finished up powering one of Britain's most famous sports cars – the XK120/140/150 series! The XK120 sports car, announced on the eve of the 1948 motor show, and originally to be built with aluminium panelling on an ash-framed body, was such a runaway success that it had to be retooled for quantity production – which was to continue until the end of 1960. The 2-litre XK100, which was announced at the same time, never went into production. The four-cylinder engine, however, had already had its brief moment of glory in Lt-Col. 'Goldie' Gardner's ex-MG record car (EX135) in September 1948.

The 3.4-litre XK engine, of course, had barely started its successful competition career when Hassan moved on to Coventry Climax, but its potentialities were already becoming obvious as it powered the brand-new XK120 to race and rally wins. One must record that XK-powered Jaguars won no less than five times at Le Mans (1951 and 1953 in 'C' Types, 1955, 1956 and 1957 in 'D' Types – the last year with a 3.8-litre engine), along with other major sports-racing victories round the world. Since 1948, the basic engine has appeared in 2.4, 2.8, 3.4, 3.8 and 4.2-litre forms in production cars, along with a few experimental 2½-litre and 3-litre racing units. The engine, which was sketched out by the Hassan/Heynes/Baily team in 1944 and 1945, was still winning races in much modified and developed form in 1963 (in the lightweight 'E' Type Jaguars), and still powers the XJ6s and Daimler Sovereigns and Limousines today, not forgetting its limited use in a Dennis ambulance and the interesting Alvis Scorpion light tank! There may be other engines which have had a production life of more than 25 years, but surely none which has graced such excellent cars and won so many racing and rally trophies along the way.

5
A fire-pump that wins races

Moving into Coventry Climax at Widdrington Road from Jaguar in Foleshill was a big step in every respect. Climax's main products were fork-lift trucks, industrial engines for many different purposes – stationary and mobile – and fire pumps. The gas turbine project already mentioned was about to be delivered to the Ministry, having been completed. I was to be Chief Engineer of this vigorous family-owned concern, and was really looking forward to broadening my experience. At first I did not particularly relish the prospect of working on fork-lift trucks, which were completely different from the fast cars I had lived with for 30 years; the main attraction to me was the engines and the new work which Mr Lee had in mind. However, almost as soon as I was properly settled I realised that there was a lot of good sound engineering – mechanical, hydraulic, electrical, and of course work on the engines – and I soon found that there were real and exceedingly interesting problems.

The company's new fork-lift truck was in serious trouble in its transmission development. The transmission of fork-lift trucks is usually the achilles heel of the design because of the nature of their work. They are heavy, move very slowly, and make a large number of starts and stops. Usually the clutch is the weakest spot, and although the trick is usually to fit large clutches with thick linings, transmission rebuilds were far too regularly required. The new truck had a very complicated transmission layout, including worm drive to the rear wheels, and any clutch change involved splitting the engine from the gearbox and lifting the engine out.

It seemed to me that we needed to take a new look at the whole problem, and as far as I could see the expertise was not in the company at that moment. I have always maintained that it is better to know where to acquire specialised knowledge than to expect to have it oneself. I make no bones about this – I always ask the experts, then make my decision as a result of judgment and experience. This was exactly what I had to do in this case, knowing that we were in dire trouble with the worm drive. I told Mr Lee that I was by no means a gear expert, but that I would discuss the problem with people like Dr Merritt and Dr Walker of David Brown Industries. This done, I went boldly into the very simple but effective changes which considerably stiffened up the worm wheel and cut down deflections to a minimum. But the major service problem remained.

After the first few months I became aware that my entry into the company had put several noses out of joint! The design office was virtually divided into fork truck, engine and fire pump divisions, each under a Chief Designer and each operating separately. I suppose each one of these Chief Designers had aspirations to become head man, and when I arrived from Jaguar they found their hopes dashed. Incidentally, very soon I was invited to join the board as Technical Director.

Although they were quite nice about this, it was clear that they were not happy. The fork truck designer was eventually offered a position as Chief Engineer at another firm, which he accepted, and our engine man moved over to Mr Lee's Iso-Speedic company, which made governors for all manner of engines. It rapidly became obvious that I needed to build up my own engineering staff and my thoughts travelled immediately to my old friend Harry Mundy.

Harry had worked at ERA on engine design in the late 1930s, then (following the closure of ERA's racing programme) had returned to Coventry to Morris Engines. During the war he had risen to the rank of Wing Commander in the RAF, then started design of the V-16 BRM engine, first at Peter Berthon's house in London, and later by setting up and running the design office of BRM in the converted maltings at Bourne. When I decided to move on from Jaguar to Coventry Climax, I suggested that he should apply for my old job, as I knew he was no longer content at Bourne. His pay was by no means outstanding there, and the V-16's development was proving difficult. The company was under-financed, which meant that all work to make the BRM raceworthy took a lot longer than it should. In the event, Harry went to talk to Bill Heynes, but decided not to join Jaguar, and was therefore still 'on the market' when I called him up to see us at Coventry Climax.

I believe I said something like: "Look, I've already got the feel for this firm, and although I cannot give you a title at the moment, come and join us to help build up our engineering strength." We met with Leonard Lee one Saturday morning, and Mr Lee offered him a job. Harry was a bit crestfallen when he wasn't offered any more money than he had been earning at BRM, but I swiftly kicked his leg under the table before he could refuse. After a bit of thought Harry agreed, and as soon as possible was installed with his drawing board, but without a title, in a little office, which was really little bigger than a broom cupboard, under the stairs.

His first assignment was to design a new fork-lift truck transmission to overcome the clutch change problems. This we achieved by arranging for the transmission drive-shaft to be capable of being pulled through the front of the box, which then allowed the clutch to be changed merely by lifting it out through the top of its housing without parting the transmission from the engine. This all took time, several months in fact, and by the time we had settled it to our satisfaction it was the end of 1950 and other great projects were afoot.

At this point I should explain that Coventry Climax had a fair tradition of making portable fire pumps and generators for the Home Office, the fire services, and the armed forces. This meant providing a light, powerful engine, driving a centrifugal water pump, all mounted in a portable framework or on a light trailer, which could be used for fire-fighting where big appliances could not reach the blaze, or as a generator set for producing electricity a long way from normal sources. This had all started way back in the 1930s, when Coventry Climax's business of supplying complete engines to car companies like Triumph, Crossley, Morgan and Swift had begun to tail off. To replace this loss of trade, Mr Lee had started to supply engines of the old Swift design to several firms who were already building motor-driven pumps and generators for the Government. It wasn't long before Mr Lee realised that these companies were taking more profit out of the deals than he was, and the next logical step was to tender direct for the business himself as the main contractor. His problems then really started, for when his tenders were accepted, he had to set his design team to deal with things about which they knew nothing. This might never have been resolved if Mr Lee had not gone over to America, borrowed an engineer from the Hale Fire Pump Company, with whom he was very friendly, and put him into the Coventry factory. The result was the reliable old unit which put out most of the fires in Britain during the Second World War.

The old 800cc side-valve Swift engine, in cast-iron and producing no more than 20 bhp with a struggle, was still being made when I joined the firm, but even its guaranteed performance was a bit borderline, and depended on the engines being in tip-top condition. There was also another much more sophisticated generator set being made the design of which was closely controlled by the Ministry people, who had insisted on a four-cylinder *air-cooled* unit of a very low power output, because they thought this would be reliable and long-lasting. That may have been so, but the problem was that they only ordered a few every year – perhaps twenty or thirty – and the orders were always accompanied by a long list of service complaints. To be frank they were a nuisance to make and were not really profitable.

However, in 1950 there was a big war scare due to the start of the fighting in Korea, and as part of a general re-equipment programme the Government decided it needed a new design of fire-pump. They were not content merely with a new pump that would perform rather better than the old one, but demanded twice the pumping capacity from an assembly weighing only half as much as the old one – in other words a power/weight ratio about four times that of the old Swift design!

Mr Lee called me in to see him one day in September 1950 to tell me the new specification. He wanted to know if we could tackle it, and whether we could meet the performance target, and I replied that I was sure we could, but I was a little anxious about the time allowed to achieve it. There was no doubt at all in my mind that we would need a brand-new and very efficient engine, and I was fairly sure that it would have to be of advanced design. Mr Lee suggested that we should go ahead as quickly as possible to beat the technical requirements, and hope that our engineering experience would allow us to keep costs down.

Right from the start of this exercise, Harry and I realised that we were faced with the same set of problems which would be met by racing designers in that we needed efficiency, light weight, and a good power/weight ratio. The specification indicated that we needed 35 bhp at around 3,500 rpm, and for fire-pump purposes continuous high-speed operation would be normal. In 1950 10 bhp per litre per 1,000 rpm was quite an advanced target, but we thought we could do the trick with a one-litre design. It was also obvious that we would need an aluminium cylinder-block and cylinder-head. A simple overhead-valve arrangement would not be adequate, especially as we would have to build in what I call a 'factor of ignorance' because there was so little time allowed that we had to 'hit the jackpot' without any development period and go straight on to fulfil the terms of the contract.

We spent some time considering our problem. In those days motorcycle engines were considered to be well ahead of car engines in technical excellence and performance. We therefore had a good look at the various designs then in being, and came to the conclusion that the general layout of the twin-cylinder air-cooled Sunbeam engine had much to commend it, particularly the wedge-shaped combustion chamber with inlet ports sloping so neatly into the chamber. Wedge heads of sorts were becoming quite popular in the United States, but originally Harry had gained experience of them on the 4.3-litre Jamieson engine for ERA in the 1930s.

We decided to adopt this layout, and also to use the same type of tappets and biscuit adjustment as used by Jaguar on the XK engine. This design originated at Ballot before the 1914-18 war in racing engines, and has been used by many other companies since. Harry schemed the engine in little over three weeks, but because of the tight schedule imposed we decided to farm out some of the detailing to save time. We lacked the sheer manpower in the drawing office to tackle the vast amount of drafting, so we enlisted the assistance of the research company which Leslie Johnson had acquired from ERA. The company initials now stood for Engineering Research Association, and it was based at Dunstable, but to all intents and purposes it was the successor to Humphrey Cook's little company at Bourne for whom I had worked briefly in 1936/37.

The next problem was to get the components manufactured, and here we were fortunate in having a very knowledgeable Purchase Director in the able Miss Norah Morris. Miss Morris had grown up with the company, advancing through stages to become personal secretary to Mr Lee before taking over the purchasing function. Although she retained her maiden name in business, she was really Mrs Magson – her husband was Chief Inspector at Coventry Climax. Happily, Coventry Climax was always well equipped with machinery, so that we could finish off anything, although we never had a foundry and relied on outside specialists for castings and forgings.

With a lot of help from the shop floor and the machine shops the first engine was ready to run in April 1951. But as luck would have it that didn't happen until a Friday afternoon, and by the time all the plumbing and mounting-up was completed by our fitters it was lunchtime on Saturday. Frankly, neither Harry nor I could wait to see if we had done the trick, so I asked him if he would mind staying on for Saturday afternoon rather than going home? (At this time he still had a home in Bourne, and only went home at weekends.) Of course, he agreed and we armed ourselves with enough spare carburettor jets and chokes to make adjustments, and set out to give our little FW engine (FW stood for 'featherweight' and by comparison with the old Swift engine it was!) its first test runs. With a very important contract at stake, we were not at all surprised when Mr and Mrs Lee came along to see what happened. We spent the first half-hour in gentle running and making sure that the settings were about right, then I took my courage in my hands and opened it right up! All this time Mr Lee had been pacing up and down the yard outside in a very agitated manner, occasionally popping his head round the door to see what all the delay was about. However, on the very first power test, without any messing about, the engine produced 37 or 38 bhp, which was well above specification. I hurriedly closed it down, and made a victorious 'thumbs up' sign out of the window to Mr Lee, who looked very relieved to find that I hadn't blown it up!

What happened then was quite unforgettable, even if it was a bit incongruous at the time. No sooner had we made sure of our 37 or 38 bhp in that funny little experimental test bed, and shut off for a time, than Mrs Lee marched in triumphantly bearing a beautiful silver tray, on which was a silver teapot and some rather delicate cups. So we all sat down on boxes or wherever there was any space, and solemnly toasted the success of the design in tea. It was a great relief to know that our theories had been proved out, and very satisfying to know later that the engine, when married to an all-aluminium pump, passed the Home Office-type test with flying colours; the result was an initial contract for 5,000 engines – the biggest single contract the company had ever had. Mr Lee was very pleased with us and showed it in no uncertain manner. Harry still recounts with relish how Mr Lee called him into his office in all its Victorian splendour, told him what a tremendous job he had done, doubled his salary on the spot and made him Chief Designer. Harry also points out that it was the first time for many years that he had actually *had* a title!

It has sometimes been suggested that we laid down this engine with an eye to racing usage, but I can state quite definitely that this was not so. If that had been the case we certainly would never have made it such an awkward size as 1,020 cc, which was neither here nor there in terms of the racing capacity classes as they existed in 1951. The FW was a neat little unit which pleased us with its performance, but we were still not thinking about any racing applications, even after we had started to design a proper Grand Prix V-8 engine late in 1952. I tell the story of this first Coventry Climax V-8 engine in the next chapter – it was the first out-and-out racing engine we designed at Widdrington Road. Incidentally, we never actually raced the first, nor the last racing designs we tackled, but all the others enjoyed considerable success.

Once we had the FW in production, at the end of 1951, we looked around our commitments to other Ministries and decided that we should try to persuade the military people to drop their old air-cooled engine generator set and take up our new unit. We thought the fact that it was in large-scale production for another Government department would help, but in that we were disappointed. To our consternation we were told that not only did they not want our new engine, but that the old engine would shortly be phased out, certainly in the foreseeable future. They said that a new horizontally-opposed two-stroke was being developed for them by HRD (the motorcycle people) but later we heard through the grapevine that they had dropped this project, and we then discovered they were working on a four-cylinder sleeve-valve engine, of all things, which was being developed by Ricardo down at Shoreham. We could see no advantage in sleeve-valve engines any more because they would be no more effective than conventional poppet-valve types, and would cost more. Their principal advantage had been the ability to maintain good performance with heavily leaded fuels. Lead deposits on valves, stems and seats had been a serious problem at one time, and a constantly-moving sleeve valve overcame this, but with the deposition problem on poppet valves now solved we couldn't see the point. In time our erstwhile customers found that this design *was* too expensive, and too difficult to manufacture, so they decided that perhaps it really wasn't what they wanted after all. By then our FW engine was becoming obsolete, and we had developed and productionised a new and smaller unit, which was even lighter while developing the requisite power.

We tried to sell this new concept of an ultra-light fire pump to other countries. We knew that people all over the world ought to be looking at this set, because it was so compact and efficient. Two men could carry a two-squirt pump quite easily over rough ground and through a building before starting to use it. It really was 'featherweight' – lighter than anything produced before. Since we knew the Americans were still using two-stroke petrol pumps with heavy bronze castings we thought there must be a good market for the FWP over there, but though we made visits and penetrated into the depths of the Pentagon we only supplied a few prototypes, and a few high-pressure pumps for one company based in Chicago. A major problem at the time was one of manufacture; our facilities were limited, and we could not afford to expand them, so although we tried to sell, we were unable to accept orders of any great size – a very contradictory and frustrating situation!

At first we had one other difficulty in the Pentagon – nobody knew anything about the company. When we said who we were the standard response was: "Oh yes, aren't you the people who make windmills?"; they only knew a Climax agricultural pump which was powered by windmills! The other common mistake was to connect our name with a company making rock drills, and no-one seemed to know us as engine designers and builders. It is here that one of the remarkable benefits of motor racing becomes apparent. A few years later, when our FPF engines were winning Grand Prix races all over the world, we had occasion to visit the Pentagon again, and on this occasion there was no doubt at all that everyone had heard of us. The difference in attitude and in knowledge of our activities was all due to the excellent publicity we had gained in motor racing, and Mr Lee was well aware of this. Indeed, during a frustrating time when we were trying to convince the Ministry to take supplies of our new engine, Mr Lee suggested that we take up racing. "After all", he said, "this may be a very good way to convince them. Jaguar have just won at Le Mans, and they have been given a contract to develop a tank engine!"

Eventually we were persuaded into motor racing more by chance than because of a definite management decision. Coventry Climax had several engines with marine applications, and it was with such business in mind that they always took a stand in the marine section of the Earls Court Motor Show. It was on this stand that members of the racing fraternity became aware of the FW, and of design of the FPE V-8, then being started, and we began to receive requests from people like Colin Chapman, John Cooper and Cyril Kieft for a racing version. The popular small sports car classes in 1953 were for 1,100 cc or 1,500 cc cars and they seemed to produce cars with much-modified Ford or MG power units. There was no purpose-built small racing engine on the market while the production units used were expensive to modify and none too reliable, either.

Quite soon Mr Lee agreed that we might have a go at race-tuning the FW engine, though he was not prepared to let it interfere with the mass of industrial work which I had in hand. At around this time we were absolutely up to our necks in the re-design of a series of high-speed diesels for various applications, and we were also deeply concerned with the fortunes of the FPE V-8 Grand Prix unit already mentioned. However, both Harry Mundy and I were well aware of the potential of the little 'featherweight' engine and thought we could convert it for automobile application in a straightforward manner. Thus the FW became the FWA, with new bore, pistons, steel crank, twin carburettors plus revised valves and porting, and was ready to race within months.

Although I had kept in touch with motor sport people whilst with Jaguar, I didn't know everyone in the business by any means. When I was first approached by Colin Chapman (who was only then starting up in motor racing) I had never met him, or even heard of him, and when I asked Harry about him, he could only say that he had read the name in magazines, and that he had started making a few kit cars. We got to know each other much better in the next few years!

Developing the original fire-pump engine into a sports-racing unit was very straightforward because we started with the advantage of a light-alloy cylinder block and head, a single-overhead-camshaft layout and a combustion chamber which was capable of great deeds if developed carefully. By increasing the cylinder bore to reach the 1,100 cc limit, raising the compression ratio and attending to camshaft timing, breathing and valve sizes we were able to double the engine's power without too much difficulty. The first engines were ready during 1954, and although we had had virtually no testing in cars, we were just in time to have an FWA fitted to a Kieft for the Le Mans 24 Hours race in June. I was so worried about its running in an untried form that Harry Mundy went over to Le Mans to help Kieft if they needed it. After all, we had designed it for use as a stationary device, and in modifying it had had to take into account all the acceleration, braking and cornering forces which would cause the oil to swish about in the sump.

However, Harry's first problem was that he could not, at first, find the Kieft team, and when they finally arrived he had to fix them up with a garage in the town to carry out work on the car before race day.

Kieft's mechanic admitted that the car had not been run at all – we had seen it only a week previously in Wolverhampton in an unfinished state – and there were many details which had to be sorted out before it could practise at any sort of speed. Apart from the sloping windscreen, which was so close to the driver's head that it threatened to cut his head off, and a space-frame built so tightly round the feet that it was nearly impossible to get to the pedals, the chief mechanic fell over one of the pit barriers and broke his arm! This immediately made Harry Mundy a *de facto* chief mechanic – something he hadn't banked upon when he went over there. Loss of oil through the rear main bearing was also a great problem at first, because there was only a simple scroll in the crankshaft bearing, and Harry co-opted Charles Goodacre to help him modify the baffling. He even rang me up on the Friday and asked if he should recommend withdrawing the car, but after a long talk in which we discussed all the work he had done I urged him to let the car start; the big worry was whether the Kieft could complete its minimum regulation distance between replenishments before it had either lost all the oil or destroyed the bearings, but we decided to take a chance and find out.

There isn't a fairy-tale ending, I'm afraid. Alan Rippon and Bill Black drove very well for more than ten hours and were leading their class, and I'm happy to record that the last-minute change to sump baffles solved the oil-loss problems, but the car was forced out by back-axle failure. The engine ran like a clock throughout, which gave us all confidence for the future. It wasn't long before our hopes were rewarded. In the Dundrod TT, later in the summer, when two Kiefts and a Lotus raced with our engine, the 1,100 cc class was taken by Ferguson and Rippon in their Kieft.

Soon our FWA, as we called it (the 'A' standing for automobile), was winning races all over the world in Coopers, Lotuses, Kiefts and several other makes of sports-racing car, and at the Widdrington Road factory we were faced with supply problems which were quite new. Usually our sales were in fairly regular quantities to industrial and Government customers, but here was a new slant on business which involved individual deliveries to small concerns making their own racing cars. Geoff Densham took on the task of racing-engines sales manager, and his famous little 'black notebooks' now contain invaluable information for historians on the engines supplied to teams from 1954 until we delivered our last racing FPF engine in 1966.

At first we concentrated on the 1,100 cc 'standard' racing unit. As outside suppliers we had to be very careful about not having favourites to whom we would supply extra-special engines. This meant that as long as 'Joe Bloggs' had the right sort of money, he was entitled to the same engine, to the same specification, as those supplied to Colin Chapman, John Cooper or Stirling Moss. Naturally there was always someone who wanted to have the 'special' which they seemed to think we just had to be sending to someone else, but in the main we did a good job of building production-line racing engines which all produced, within a very little, the same power curves, and kept everyone happy. This also meant that providing an engine was kept in good condition, victory or defeat in any race was almost certainly due to the quality of the driving or the virtues of the rest of the car.

Of course there was bound to be someone who thought he knew better than we did, and who would buy one of our engines, take it round to a specialist tuner, and try to find more power than we were providing. Alf Francis, in his very interesting book (Alf Francis, Racing Mechanic) tells a lengthy story of work done on an engine for Stirling Moss, and it may not be entirely coincidental that Moss had quite a lot of trouble with his Cooper-Climax sports car that year! That particular engine had been taken round to Barwell Engineering (founded by people who had learned their craft with Harry Weslake), who then applied their usual dodges of opening-up the ports, fitting bigger valves and so on. They didn't realise, or perhaps wouldn't accept, the fact that Coventry Climax were selling a purpose-built engine in which ports, valves and details were already correct, and the result was trouble! Stirling Moss always wanted something more than the others, whereas he could certainly win races with less than his rivals. Yet he must have spent a year on this engine, and had a lot of trouble. In the end he came back to me and said: "What's the trouble? Why can't I win races like the others?". I said that he should send the engine back to us, we would return it to the proper, standard condition, and then he might start to win! When Stirling wanted to know how much it would cost I said: "Something like £120", and he nearly threw a fit. It wasn't until later that I discovered he had spent between £200 and £300 on it already. After that we used to have an expletive going round the workshops about 'Barwell and b_____ 'em' . . .

On another occasion, Colin Chapman, who was always sharp as a razor but charming with it, sent back an engine for test before he fitted it to a Lotus for a particular sports car race in Britain. We asked him what he had done to it to make it worth re-testing, and when he said: "Nothing at all" quite cheerfully, we had no hesitation in giving it the full treatment. The result, which I think we anticipated, was that the engine 'blew up' on the test bed, and when it was stripped we found (as we had suspected) that among other things the compression ratio had been pushed up a long way. On this occasion we felt that we had to make a stand over the deception, so we sent Colin's engine back in its blown-up state, and let him know that his ploys had been rumbled. Then I got on the 'phone to his big rival, John Cooper, and asked him if he would like to try an experimental unit we were developing – for the same race that weekend! I'm happy to say that we proved our point by Salvadori winning that race. After that we had to introduce a further stage of tune, so as to maintain our policy of fair do's for everyone!

Naturally there was always a demand for more power, and after we had offered a conversion kit to uprate the existing 1,100 cc FWA we were persuaded to enlarge a few units to make cars eligible for the 1,500 cc class. This we did, and with considerable success, but as an engineer in preference to a racing enthusiast I was never very happy about the 'stretching' business. We had to go in for this more and more in later years, but it was only because we were all ordinary fallible engineers who had built in that little bit of extra space, with strong scantlings and the previously mentioned 'factor of ignorance', that I think we got away with it! The FWA therefore became the FWB, with 1,460 cc, and apart from winning sports car races with it, Cooper installed FWBs in a handful of single-seaters which were really prototypes for the new 1½-litre Formula Two. The FWB Coopers were built in 1956, Salvadori winning the first race for them at the British Grand Prix meeting at Silverstone, in July, when he lapped only slightly slower than the full-size Grand Prix cars. We also decided to turn the racing modifications to good advantage by developing an uprated fire-pump which used FWBs. The pump version was FWBP, and could help pump 500 gallons every minute instead of the 350 gallons a minute originally required; this was virtually 'for free' as there was no real weight penalty.

These engines sparked off great enthusiasm both inside and outside Coventry Climax for further racing design, and I must stress again that such enterprise simply would not have been possible without the help and encouragement and burning enthusiasm of Mr Lee. Not that the FWA and FWB engines we sold made a loss for the company. I have an idea that they more than covered their costs, and because much of the engine could be machined on the same jigs set up to deal with the steady flow of orders for fire-pump engines, they were not so much of a disruptive element in the factory as the Grand Prix engines became later.

Perhaps I should also record that no fewer than 697 of the 1,100 cc 'racing fire-pump' engines were made eventually, which was several times more than any other engine sold until then exclusively for racing, and is still probably a record of sorts today. We were reluctant to make many of the enlarged 1,460 cc engines because of a distrust over reliability, and only 35 were built in all.

However, the impressive record was surpassed by a variation on the theme which was inspired by Colin Chapman. Chapman, who had been making small numbers of sports and racing cars for some years, decided to make his first true road car, which arrived in 1957 as the Elite. He is a very persuasive fellow, and managed to convince Coventry Climax that a special 'productionised' version of the FWA should be built to power his Elite, and guaranteed that at least a thousand would be sold. This engine was the FWE (E for Elite, of course) and was a combination of FWA and FWB in that we used the FWB's block and cylinder bore allied to the shorter FWA stroke. At one time we discussed the possibility of producing cast-iron cylinder blocks in place of aluminium alloy to bring down costs, but we finally abandoned the idea because we could not cope with further relatively low-volume cast-iron work. In single-carburettor form, the FWE produced a very unstressed 72 bhp, with great built-in potential for race-tuning. We didn't start to make these in quantity until 1959, by which time racing FWA production was tailing off, but the Elite wasn't withdrawn until 1963 after the thousandth engine had been built. One thing led to another, and we were approached by Jack Brabham, who had his own ideas about Coventry Climax-engined road cars. In Jack's case this comprised a major rebuild of cars like the Triumph Herald and the MG Midget, with the FWEs slotted in in place of the standard units.

All this exposure to the motoring enthusiast's market was pure joy to our advertising manager, of course. Previously it had been quite difficult to get the name and image of Coventry Climax across in anything but the most mundane way, but soon after Cooper and Lotus started to use our products in their racing cars the obvious happened – some bright spark invented the phrase which stuck with us through the next ten years. Our well-known products became 'The fire pump that wins races', and this slogan was usually accompanied by a priceless sketch of a bravely-driven fork-lift truck chasing a fire pump trolley chasing a Cooper racing car around a corner, obviously on a race track because it was festooned with advertising banners. All very colourful, and I am glad I was never there to see an actual occurrence of it!

But as far as motor racing was concerned, we were already well committed. 'The fire pump that wins races' was only an extensive conversion of an existing design, but Mr Lee had more important projects in mind. Even before we had started the automotive conversion of the FW engine, he had instructed us to start designing a proper racing engine. There were to be no compromises, and no enforced use of existing parts. What he had in mind was nothing less than a new engine for the 1954 2½-litre Grand Prix formula, and he wanted a world-beater. For a time it would have to be done as a spare-time project, but to Harry Mundy and myself this was no great deterrent. We had been asked, and actively encouraged, to do something that every engine designer enjoys – to build a racing engine without compromise, and with only a single limitation of cubic capacity. We could sacrifice a lot of evenings and weekends for that.

Coventry Climax FW single-cam engines

The original FW fire-pump engine produced a very under-stressed 36 bhp at 3,500 rpm, as the camshaft design and general tuning was very conservative for its endurance usage. Following the decision to convert the engine into an 1,100 cc competition unit, a prototype was built having the very odd 1,020 cc capacity (but as Hassan says: "For the fire pump light weight and pumping capacity was vital, it didn't matter what size the engine was so long as it was economical enough!"), and this immediately produced a very creditable 64 bhp at 6,000 rpm on a compression ratio of only 8.8 to 1. The main structural change was to a forged-steel crankshaft instead of the standard cast-iron item, along with twin SU carburettors, free-flow manifolding and a suitably ambitious camshaft timing.

The cylinder-head layout, with a wedge-head combustion chamber and a line of inclined valves under the single overhead camshaft operating bucket tappets, was one with which Harry Mundy had been familiar since 1938, when it was used at ERA by Murray Jamieson for his 4.3-litre production engine – a project cancelled following his unfortunate death at Brooklands later that year. Both Mundy and Hassan remained firmly convinced that a single-cam design went more than half-way towards the then ideal of a twin-cam head. Fifteen years later they found that they could even make it do as fine a job as a twin-cam, which is one reason for the Jaguar XJ12 unit only having one camshaft above each bank of cylinders.

To reach the 1,100 cc capacity limit of an increasingly important sports car class involved nothing more than a simple bore increase, the same stroke and indeed the same connecting rods being retained. The definitive FWA engine had a bore and stroke of 72.4 mm and 66.7 mm for a swept volume of 1,098 cc. The first engines, like that used in the Kieft at Le Mans in 1954, had a compression ratio of 8.8, and boasted a power output of 72 bhp at 6,100 rpm, but by increasing the compression to 9.8 for the 'production' racing engines this was raised to 75 bhp at 6,500 rpm, allied to a very creditable peak bmep of 160 lb/in^2.

It didn't take long for Coopers and Lotuses to corner the market in the 1,100 cc sports-racing business, and when they were joined by the very advanced front-engined Lola the FW engine became completely dominant. Cars which had previously won races with super-tuned MG XPAGs of no great reliability, or with modified Lea-Francis engines of pre-war design, were almost instantly obsolete. The problem, of course, was that as soon as all the top drivers had use of the same cars with near-identical engines, someone had to start asking for more, or – worse – trying to tune the engines further. Hassan has already mentioned Stirling Moss' attempts to make extra-special units which in the end had to be rectified at no little cost by Coventry Climax; there were many others, and few produced worthwhile *improvements!*

Coventry Climax, however, developed the engine further, and with the aid of a revised camshaft profile and modified inlet manifolds produced a Mark II version churning out 83 bhp at 6,800 rpm. However, by the end of 1955, following a season in which the cars seemed to win everywhere except at Le Mans, there were demands for more, both in terms of power and in deliveries. Seven 'prototype' FWAs had been delivered in 1954, followed by a batch of 100 engines which sold for the princely sum of £250 each! An even bigger batch, to be built among the much greater number of FWs for pumping duty, was laid down for 1956, along with a series of 24 FWB engines with a capacity of 1,460 cc. The 'B' in FWB meant nothing in particular (Hassan: "B was the next letter after A so we used it") but the engine itself was meant to make the little Lotus Elevens and Manx-tailed Coopers even quicker, perhaps into outright race winners on certain tracks.

The FWB's 1,460 cc was achieved by increasing both bore and stroke to their practical limit – to 76.2 mm × 80 mm. The revised engine was therefore 'under-square' – then, as now, a rather undesirable state of affairs, and one certain to limit revving capabilities – and as the cylinder-head and manifolding was the same as before it was not surprising that the peak power rose to only 100 bhp at 6,000 rpm. There was a marginal increase in peak torque, which was now developed much further down the scale, but the engine was never considered other than a stop-gap to keep the constructors happy.

No fewer than six of the new 1,460 cc engines appeared at Oulton Park for the British Empire Trophy race in April 1956, where they set fastest practice times outright (faster, for instance, than the 2½-litre Aston Martin DB3S of Reg Parnell and Ron Flockhart's Ecurie Ecosse 'D' Type Jaguar). In the race, the biggest Maserati of Benoit Musy was quicker in its heat, but in the final, where the 1½-litre cars were given a handicap start of 40 seconds, they could not be caught. Indeed, Stirling Moss' bob-tail Cooper led five other 1½-litre Coventry Climax-engined Lotuses and Coopers home ahead of all other cars in the race. More important was that the fastest 'other make', Ron Flockhart's Jaguar, finished more than one minute adrift, so the handicap was only academic anyway; the 1½-litre FWB was indeed a very satisfactory stop-gap.

By the end of the summer John Cooper's little Surbiton factory had built their first 'proper' single-seater racing car – ie, one which was not propelled by a motorcycle engine. This was the FWB-engined Formula Two prototype which Roy Salvadori and Tony Brooks used to such good effect for the rest of the season. It was also a splendid way of shaking out all the teething troubles of such an installation before the very first twin-cam Coventry Climax engine (the FPF) could be made available early in 1957.

However, the short-lived FWB successes were not the end of the road for this first Hassan-Mundy enterprise. The effervescent Colin Chapman was busily designing the stressed-skin glassfibre-structure Elite, which was to have the FWE version of the same engine with a swept volume of 1,216 cc from a bore and stroke of 76.2 mm × 66.7 mm – a true hybrid. To emphasise the effect of a torquey production-car engine, the basic unit had only a single SU carburettor (72 bhp at 6,100 rpm) but a slightly detuned sports-racing version with twin SUs gave 83 bhp at 6,300 rpm. Both units had a compression ratio of 10.0 to 1, and virtually all components were from the race-proved FWA. One can emphasise the featherweight characteristic by quoting the weight of this road-equipped engine, which was a mere 215 lb in twin-carburettor form.

The Elite took ages to bring into quantity production (it was 1959 before the line was moving properly), but 1,000 were made before it ran out in favour of the Elan in 1963. The FWE, therefore, became the most common FW variant, for apart from these 1,000 engines about 150 went to Jack Brabham for his engine-swops into Triumph Heralds, Austin-Healey Sprites and the like, and many more were supplied in smaller quantities; total FWE production was 1,355 units.

But from FWA to FWB and then to FWE - what happened to FWC and FWD? FWD is easily explained - it was a diesel version for possible outboard use, of which only one was made, but FWC was a very special animal indeed. It was a one-off, made for the ever-persuasive Colin Chapman for Le Mans in 1957. The car it graced was a Lotus Eleven, driven by Allison and Hall, and its capacity was only 744 cc. The FWC was simply an ultra-short-stroke version of the FWA, built specifically for Lotus to win the Index of Performance. This was a feat they achieved easily - the car finished a rousing fourteenth and averaged 90 mph - for which the French, who thought the Index was their own property, never forgave Chapman. To reduce the capacity to less than 750 cc a stroke of 45.2 mm was needed. Naturally, the cylinder-head's breathing arrangements were almost completely unsuited to this, with the result that the maximum power of 59 bhp was developed at 8,000 rpm and there was virtually no driveable power below 6,000 rpm. Yet it worked, and worked well, but the engine was only used twice. The second time at Le Mans its driver disgraced himself by getting stuck in the Mulsanne sand for a couple of hours, so there was no repeat of the 1957 triumph.

The FW series was both the first of all the successful Coventry Climax competition engines and the most prolific. A total of 1,988 of all types were delivered for automotive use, in addition to the several thousands of basic fire-pump FWs which were built for Ministry contracts.

6
Never believe the opposition

Leonard Lee had two great enthusiasms. In business he was never happier than when he was involved in new engine designs, and he was a great patriot who liked to see the country performing well in motor sport. He was the inspiration in getting Coventry Climax started on the road to Grand Prix success in the early 1950s. Of course, Harry Mundy and I were fully aware of the situation in motor racing, where there were several keen and capable teams who all suffered from the lack of really powerful engines, and we talked about the studied contemporary racing trends, but the idea that we might be able to design racing engines at Coventry Climax never really entered our heads. At the time, too, we were far too busy with the featherweight engine and its derivatives in fork-lift truck problems, and in updating the various marine and diesel engines which the company was making for its customers. In Grand Prix racing the situation was rather desperate. The BRM project, in which Harry Mundy had been closely involved, was staggering from crisis to crisis. The car's road-holding was none too brilliant, reliability was difficult to achieve, and the supercharged V-16 engine had a rather singular power curve – with very little in the middle range and steeply rising power at the top end. This is not conducive to high lap speeds except on really high-speed circuits where cornering and acceleration are of small consequence and a high maximum speed is of great importance.

In 1952 this situation changed. The Grand Prix formula was to be for $2\frac{1}{2}$-litre unsupercharged engines from January 1st 1954. There was an alternative formula for supercharged 750 cc engines, but since Ferrari had already proved that their 'unblown' engines were very competitive no-one took it seriously. Although this might make it easier for new British engines to be competitive, there was still no guarantee that this would happen. There wasn't really any future for the heavily modified type of production engine any more, and although the four-cylinder Alta was a good conventional design, it would certainly not be powerful enough against the Ferraris and Maseratis.

Although Coventry Climax were not involved in motor sport, quite a number of enthusiasts knew that Harry and I worked for Mr Lee. Nevertheless, the first we heard of requests to the company to start designing racing engines came from Mr Lee himself. It all started when the bosses of companies like Cooper, Connaught, HWM and Kieft made a joint approach to him. Their plea was that there was no independent company undertaking the design of a Grand Prix engine. Would he be willing to authorise a design from Coventry Climax, to be tackled by Harry and myself? In the event, it was at the annual Shell oil company's Motor Show party in the autumn of 1952 that Mr Lee was finally convinced, and he instructed me to go ahead with the layout of a new engine for the 2½-litre formula!

Harry and I were overjoyed by this, and keen to get on with studies at once. However, I knew that there was little room for manoeuvre in our programme of 'bread and butter' work, and it would be out of the question for Harry to be involved on a full-time basis. We had about fifteen months before the start of 2½-litre racing, but I knew that by the time we had done a 'spare-time' design job, got the first prototype running, sorted out changes and improvements which inevitably would be needed, and made enough engines to supply the teams, the time would be easily swallowed up.

We had no difficulty in agreeing on a basic layout. We estimated that we would have to achieve not less than 250 bhp (100 bhp per litre) right away, and that there would have to be room for development. In our view, we could not get this sort of instant power without using high revs and a large piston area. We never seriously considered a four-cylinder layout (1959 was still a long way away – if only we had had a crystal ball!), and we thought a straight 'six' would be too heavy, while a straight 'eight' would be too long, heavy, and difficult to keep free from crankshaft torsional vibrations. Our solution, logically argued, was to have a V-8 with rather 'over-square' cylinder dimensions and simple, rugged construction. Well over ten years after that it was very satisfying to know that Keith Duckworth reached the same conclusions when he was commissioned to design the DFV engine for Ford.

Harry and I discussed the design in detail, in the light of all our combined experience, which embraced engines as diverse as the V-16 BRM and the Jaguar XK, and drawing began. As Harry was Chief Designer and I was Technical Director, it was inevitable that the actual drawing and design was done by Harry, while I carried out the duties peculiar to my position. We would discuss his drawings for some time, at least twice a day, so that work progressed under close scrutiny and became an amalgam of our ideas. Because our normal commercial work also had to be dealt with, Harry had to put in a lot of time outside his normal working hours, and to facilitate this we arranged for him to have tea in the directors' dining room at the end of his normal working day. He would then be able to turn his thoughts to the new Grand Prix design, return to his drawing board, and slog away alone until perhaps 9.00 pm every evening.

Harry and I were, and still are, extremely good friends, a happy state of affairs that has continued unabated over many years, both as colleagues and

Colin Chapman persuaded us to build him a special road-tune version of our single-cam FW 'racing fire-pump' engine to power his first genuine passenger car, the Lotus Elite, which was announced in 1957. The FWE's 1220 cc displacement came from combining the FWA's stroke with the FWB's bore

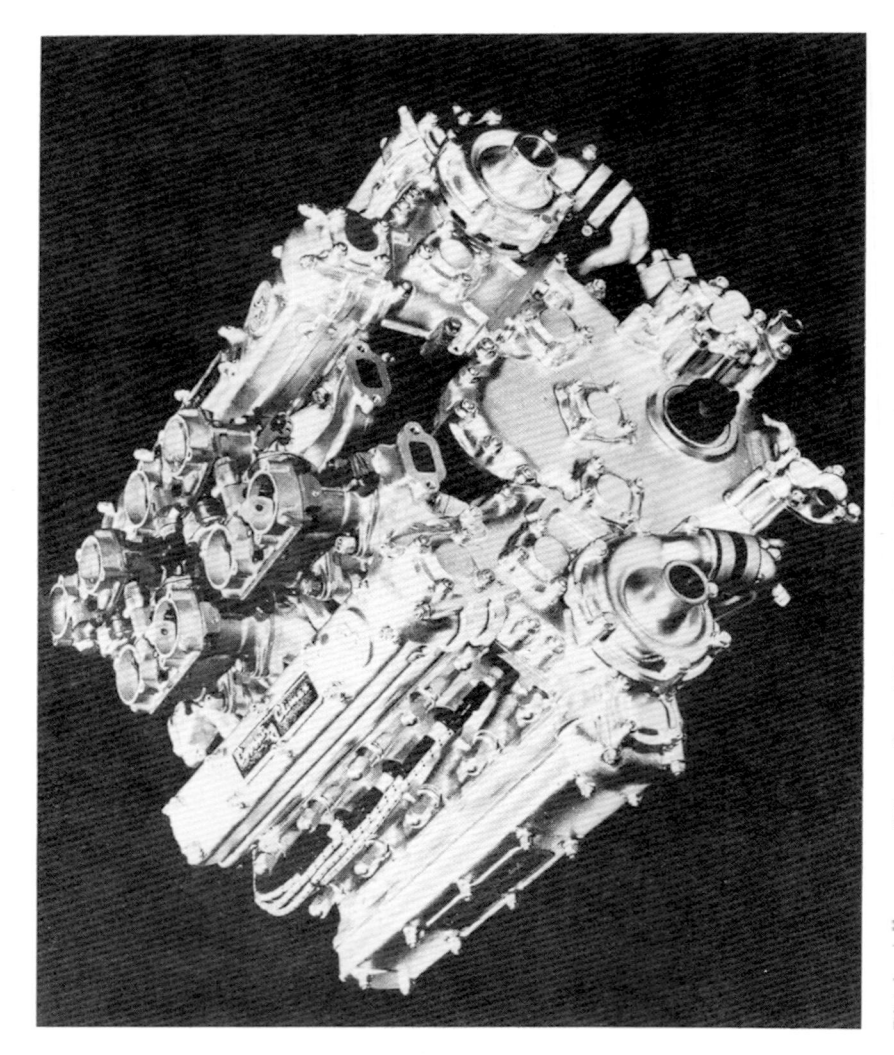

The first Coventry Climax Grand Prix engine, the 2½-litre FPE V-8, equipped with carburettors. Had it been raced it would have used fuel injection, and would certainly have been more powerful than the Mercedes-Benz straight-eights of 1954 and 1955. We designed the FPE in 1952.........

apart. I think it stems from a real appreciation of each other's qualities, a great sense of humour, a real enjoyment in discussion and argument on controversial technicalities, and a deep love of engines. Hence our twice-daily examination of progress, and the discussions which ensued, ensured that the design progressed harmoniously from both the technical and the personal points of view.

While we were being part-time racing engine designers, Ferrari and Maserati were dominating Grand Prix racing, which in 1952 and 1953 was for 2-litre cars. Their engines were supposed to give between 180 and 200 bhp (90 to 100 bhp per litre) so we felt that our projected initial output of a bare 100 bhp per litre would be only just enough. We also knew, though not with much detail, that Mercedes-Benz were planning to return to Grand Prix racing in 1954. In view of the way they dominated racing in the 1930s, and swept the sports-car racing board in 1952 after years away from the tracks, we knew their cars would be quite something. Therefore we thought that we should be looking for quite a lot more power, almost as soon as we had the basic unit reliable enough to drop into a car.

It is usual for all design projects to be identified by a code, and at Coventry Climax it was always thought by management that the code should have some obvious connection with the actual project. As this particular project was one that Mr Lee was pressing himself, perhaps without the 100 per cent agreement of other members of the board, the code FPE was chosen as a rather thin attempt to camouflage the Grand Prix project with letters suggesting a Fire Pump Engine! It could also have been an insurance against the accountants' viewpoint should the project ever have failed. However, no-one from Leonard Lee down to the office boy ever seriously thought the FPE would form the basis of a sophisticated pumping set!

In spite of the limited time we could devote to the new engine, we managed to get one running in the summer of 1953. One of our main problems – as usual – was the difficulty of persuading some companies to supply one-off items for experimental development. Many an engineer would agree that it is really a lot easier to get delivery of thousands of identical components than of the one necessary to prove the design in the first place. It was very much easier to get 'one-offs' done in my Brooklands days.

Once we had the FPE running on the test bed we were satisfied that the layout was basically right. We would now admit to one or two mistakes in the design, but they were fairly easily overcome in development. We took advice from a Rolls-Royce specialist on crankshaft torsional vibrations (this man had a legendary reputation for taming torsionals on the first Merlin V-12s) and we apparently chose the worst of the alternatives in firing order which he offered us, where each bank of cylinders fired sequentially. The result was not mechanical disaster, but due to the very odd induction impulses set up in the intake manifolds, and the effect of the exhaust impulses, we witnessed the remarkable sight of fuel spraying back out of the intakes at times, and nearly reaching the test bed roof! It was all very dramatic and spectacular, but not conducive to good performance or even reasonable fuel consumption. The fire risk was also extremely high. A simple change in firing order, to a more usual sequence, cured that one, and we still didn't have a crankshaft-torsionals problem.

We had a major development headache because at first we chose to use hairpin valve springs, following the latest motorcycle racing engine practice. In all the running we carried out on the V-8s these were very troublesome and they were dropped in favour of conventional coils. In theory their advantage was in low inertial weight, which should have increased their reliability at high rpm. In the event we established that they were very difficult to manufacture and they suffered from repeated breakages due to difficulty in obtaining a good enough surface finish. Due to the spring wire being subjected to direct bending rather than torsion they were also much more susceptible to surface defects.

Vanwall, whose engines won many Grand Prix races in 1957 and 1958, persevered with hairpin valve springs and made them work. Perhaps we could have done the same, but in the face of an easy and simple alternative which worked well we were not prepared to struggle on. I have always been in favour of the simple solution to any problem, if a choice existed, no matter how unfashionable it might be. Hindsight shows us to have been correct in that no-one now uses anything but coil springs up to 12,000 rpm.

The very first day we ran the V-8 we also learned some hard facts about tappets. Instead of the very satisfactory case-hardening process on steel, which eventually helped the engines to win so many times for Coventry Climax, we originally decided to use the nitriding process. We did not realise just how brittle the nitriding treatment would make these parts, and within minutes of starting up the engine they began to break. They flexed – all tappets made in this way flex – and the cases cracked with a noise like machine gun fire.

But these were normal development problems, and as engineers to whom a working day without problems would have been rare and blissful (and we would have wondered what we were missing!) we could cope without distress. Besides, we had bigger worries. In 1954, when we should have been ready to produce several engines for customers like Cooper and Connaught to use in races, we were still worried about the engine's performance. We had started off using a set of twin-choke Solex carburettors which fitted very neatly into the vee of the engine, but they were not specifically racing instruments, and we could not get enough airflow through. Next we tried Webers, about which we knew very little in those days, and we also decided to try the SU fuel injection which was adapted from a successful aero-engine system. That was a bit new, too, and SU were learning as they went along. This was also being developed for the Alta engine which Connaught had chosen for their first Formula One car.

All the time we were pressing on with exhaust system tuning and power testing, and if only we had known then what Gray Ross discovered in 1961 about V-8 exhaust system extraction effects we might have made progress more quickly. On the FPE we never linked pipes from one side of the engine to pipes on the other side (a trick which later gave us 'power for free' on the FWMV) but we soon dropped the conventional manifold; instead we turned to separate exhaust stub pipes which can be tuned to give good results, but only within a very narrow speed band. It was also very difficult to deal with in the test shop. We were under constant fire from the production manager who regularly came into the shop and ordered immediate cessation of running because the works were threatening strike action if the noise did not stop. In fact, we were always in this sort of trouble at Widdrington Road, which was in a densely populated area of Coventry.

We kept on hearing press reports, rumours, and stories through motor racing's 'grapevine' about the fantastic engines and performances our rivals were claiming. We also knew that no matter how or where we altered things, we still ended up with a test-bed torque curve with a profile rather like the Himalayas; this was so pronounced that there were certain periods where the engine had to be nursed by throttle manipulation to accelerate up to the next step in rpm - you still have to do this on very high-output engines today, but we thought then we had made a serious mistake! We were quite convinced that cars using this engine would be just about undrivable, and probably not fast enough anyway.

But almost as soon as we had the engine going properly, with the aid of a real 'witches' brew' of an alcohol fuel supplied by Mr Rowntree of Shell, we beat our 250 bhp minimum target, and later we pushed this up to 264 bhp on SU injection. Incidentally, it was not until 1961 that we ever managed to persuade the four-cylinder derivatives - the FPFs - to deliver more than 100 bhp per litre on straight aviation fuel. By then, though, we were concentrating more on fat torque curves and leaving the peak power figure to take care of itself.

By the middle of 1954 we were somewhat depressed. Although we had only just decided to enter Grand Prix racing, didn't depend on it for our existence and were not committed to supplying the V-8 FPE against a deadline, we did not want to let anyone down. Nevertheless, we were not prepared to release the engines until we thought they could win - it would have been bad publicity for us. All the stories of phenomenal engines from Ferrari, Maserati and Mercedes-Benz convinced me that we still had a great deal to learn on racing engine design, and that we should not offer up our poor efforts to be slaughtered on the race tracks of Europe. We were very green, then, and because we had no previous side-by-side experience of our rivals we had no choice but to believe their claims. We saw no reason why world-famous and successful teams should bother to claim over-high figures. Surely victory was enough and they could state the truth?

We were fooled. Years later we learned that the best Maserati engines used by Fangio in 1954 produced a mere 230 bhp, and that the Mercedes-Benz

W196 cars he used to clinch his World Championship pushed out 256 bhp with a very peaky torque curve. If only we had known *then* we would certainly have pressed on. We know now that we had a very competitive engine in the FPE, and it would probably have been the most powerful in 1954 if only we had been courageous enough to let any go out of the gates.

The Connaught (later called the 'Syracuse' car after Tony Brooks' race victory in Sicily) made provision for the FPE V-8 engine, but Cooper never made a Formula One car to take it. When Harry and I learned that we had been misled by false claims we resolved that never again would we believe figures that we could not check out ourselves, although as we rarely had the opportunity to test rival units, we had to rely on our own judgment. For example, we knew that the FPF four-cylinder engine was way down in maximum power compared with the V-6 Ferraris, perhaps by 50 bhp, but in the hands of a Jack Brabham or a Stirling Moss, and in a nimble little mid-engined Cooper, our modest work-horse, with higher-torque, proved to be adequate, especially as it was so supremely drivable.

In all our years in the Grand Prix field, we only managed to power-test one competitor's engine, and this was only because we were supplying a concern which also used the other make. The engine in question was a 'customer version' of the $1\frac{1}{2}$-litre V-8 BRM, which proved to be competitive with but no better than the FWMVs we were supplying at the time. But even then we never took it apart to see how it ticked – there was barely enough time to put it on the test bed and check out its potential. The relatively low figures produced by early Maserati 250F engines was proved when Jaguar obligingly tried out Stirling Moss' own engine at Coventry – after a full season its maximum output was down to no more than 215 bhp or thereabouts, in which condition it was still winning races in Great Britain!

But all this is in the past, and the fact was that I recommended to Mr Lee that we should drop the Formula One project. However, the current Formula Two racing which Ferrari had dominated for several years was finally declared obsolete, and the CSI decided that there should be a new formula, for unsupercharged $1\frac{1}{2}$-litre engines, which would have to run on straight petrol. They made this announcement in 1954, but it was not due to come into effect until 1957. That gave many people plenty of time to think about their new racing cars. By this time, too, our featherweight FWA engine was beginning to win races in the sports cars. The people who were building them – Cooper, Lotus, Lola and others – decided they would like to build single-seater cars for the new formula, and started urging us to do something about a new engine for them.

We were not at all unhappy about these requests, which could not have come at a better time for us as we had decided that the V-8 FPE could be developed more cheaply and more quickly if work was done on a four-cylinder rather than an eight-cylinder version! The requests for a new $1\frac{1}{2}$-litre engine gave us an excuse for designing this four-cylinder version, using a lot of the V-8 components, or components of very similar design, including the cylinder-heads, on which we could continue development work while satisfying our customers. When we started this project we believed that all useful work completed on the four-cylinder engine would later be applied to the V-8, but once we had swung into action on the 'four' and started to win races with it, there never seemed to be time to return to the V-8, and no engines using all the accumulated experience gained on the FPF were ever built.

The new engine was much more than 'half an FPE' but the basic design parameters were the same. Using the same cylinder dimensions would have produced a 1,250 cc engine, so with some small modifications to dimensions we could easily increase to the desired limit. Fortunately the cylinder-head was adequately long enough, with sufficient space between cylinders, as is usually the case with V-8s, so we could increase the bore sufficiently. We were fairly sure that the valve gear and reciprocating parts were strong enough to look after increased loadings due to the larger capacity. During the winter of 1954 and spring of 1955 we settled down to turn our joint ideas into a robust but straightforward engine, and so the wonderfully successful FPF engine was born. Of course, we had no idea that it would be produced for nearly ten years, or that it would give rise to so many variants, and certainly at that stage I did not think it would ever be larger than the $1\frac{1}{2}$-litre size we were laying out. Why was it called an FPF? It was just the next sequence of letters after FPE, and we never put words to the letters.

At this point I feel that I ought to clear up a misunderstanding that has surfaced from time to time over the years. Coventry Climax had that very effective advertising slogan: 'The fire pump that wins races' which was perfectly valid when the engine in question was the featherweight FWA. I suppose it was logical that people should think of its successors as being fire-pump derivatives, too, and so the legend grew, and at times prospered, that the FPF also powered a fire-pump. It didn't, of course. The experience we gained over the years from racing was built into other projects from time to time, but there were no direct FPE or FPF applications in fire pumps. However, I like to think that the industrial engines, for stationary usage, boats, fork-lift trucks and military purposes, all benefited from our racing triumphs.

Unfortunately the FPF was to be the last important design on which I was to have Harry Mundy's invaluable help for the next ten years, because one day in the spring of 1955 he came to me with the news that he had been offered a new job as Technical Editor of the influential magazine *The Autocar,* and it had all come about because of his time at BRM. Harry had set up the design offices of BRM in 1946, and took on as his first assistant a man called John Cooper – not, incidentally, the John Cooper of Cooper Cars fame. Cooper later moved to *The Autocar,* becoming Sports Editor when Sammy Davis (my old Bentley Boy colleague) decided to retire. Harry admitted that he knew nothing about journalism, but didn't think it would be difficult (he still doesn't!). When I heard that he had also been offered a large salary, almost on a par with mine and much higher than nearly everyone else on the magazine, I realised that this looked like the parting of the ways. Harry saw it as a way to learn much more about racing and non-racing subjects (he had really been involved in racing car design since 1937 with only very short breaks) and eventually he decided to go. I was placed in a similar quandary to that faced by Bill Heynes in 1950. I knew I did not relish losing him, but it was obviously such a good move for him that I could only wish him well, and hope that we could keep in close touch.

However, even though it meant that I lost a very good colleague, Coventry Climax ultimately gained a great deal of favourable publicity. As an acknowledged expert in high-performance engine design, Harry proceeded to write an outstandingly complete analysis of some of the world's finest engines, and over the years his coverage of our own new products was comprehensive. Almost every one of these analyses was accompanied by superb cutaway drawings (by Vic Berris) which are now collectors' items.

Harry's departure meant that I had to reshuffle my staff. After a time I brought in Peter Windsor-Smith from the development workshops to be in charge of engine design. Peter had once specialised on diesel engines, but soon adapted to racing engine work with enthusiasm and skill. Gray Ross took over Peter's responsibilities in the development shop, while Harry Spears was responsible for the important job of assembly, overhaul and test of customer engines, plus liaison with customers. Two of Peter's young assistants who have since made their own names were Ron Burr and Hugh Reddington. Both had been apprentices in the drawing office under Harry Mundy, who regards them as his own property and proof of his own training methods. Ron Burr later moved to Lotus to design the four-valve Lotus engine now used in the latest models, but soon after its completion we persuaded him to return to Jaguar, where he became Harry's deputy on engine design. Hugh Reddington moved over from Widdrington Road to Browns Lane after the takeover, also to work in the Jaguar design offices.

The story of our successes with the FPF range of engines has been told many times, and the details of the different versions we produced are listed at the end of the chapter. I can honestly say that when we originally laid it out, it was a purpose-built $1\frac{1}{2}$-litre design specifically for the new formula. We did not consciously build in any physical stretch for future enlargements, nor over-design the main structural components in case of future uprating. The engine designer is always faced with the problem of whether or not to design for future enlargement. If a racing engine is to be really competitive it must be designed right down to a minimum weight and size to give the best results, so there is rarely any scope for increased dimensions within. Any scope that *does* become available is due to the ingenuity of the designers seeking enlargement, and also to the previously mentioned factor of ignorance which is included in the initial design due to incomplete knowledge or experience. A very good example of this was with the original featherweight FW engine, which was eventually replaced by the FWM – a potential 1,460 cc unit was replaced by an engine of 742 cc to do the same job.

The same thing happened with the FPF, which began life at 1½-litres but was finally persuaded out to 2.7-litres, still using the original size of crankcase. By the time we came to design the next generation of engines, the V-8 FWMV 1½-litre, we were much more experienced in the philosophy of minimum-size, minimum-weight design, with the result that the FWMV could only be pushed out to 2-litres by sacrificing the benefits of a short stroke. Generally speaking, the more sophisticated the design, the less likely is there any capability of enlargement. However, enlargement should not be confused with improvement!

The first enlargement was from 1½-litres to nearly 2-litres, achieved mainly by a longer stroke, but this wasn't really our idea; it was Cooper who really pressed us in the first place. However, once we had achieved the conversion we found that there was a good demand for the 2-litre, particularly for use in some sports cars. We left the cylinder-head, camshaft timing and breathing exactly as first designed, so in a way the 2-litre engine was self-limiting.

That stretch didn't make me too unhappy, but when, at the end of 1957, Cooper and Lotus came to us and pleaded for a further stretch I began to feel thoroughly uneasy. Having seen some encouraging results with the 2-litre cars, both John Cooper and Colin Chapman were convinced that they could start to win with more power. I understood their enthusiasm, and I was as keen as anyone for Coventry Climax to start winning Grand Prix races, but I was unhappy about the limitations of strength and rigidity which I was sure would show up if we pushed our luck too far. I think it is a measure of our caution that we only ever made four 2.2-litre engines – two each for Cooper and Lotus – and as an engineer I could never be proud of the way we needed a sandwich plate to support the longer liners, but I was agreeably surprised that this method stood up to a whole season's racing. Then Jack Brabham, who was as much of a mechanic as a racing driver at that time, evolved a clamping plate which helped to hold together the bottom of the crankcase and the main bearings, which were beginning to crack up under the increased inertia loading.

The 2.2-litre versions did well in 1958, better than anyone could have hoped, although ironically, the first two Coventry Climax Grand Prix wins, both by Rob Walker's Coopers in 1958, were with 2-litre engines! Inevitably, before the end of the year, John Cooper and Colin Chapman were back, pounding away at Leonard Lee to authorise yet more work on the engines. This time they had a rather strong case because it was known that Tony Vandervell was about to withdraw his very successful team of Vanwalls, thus leaving the field open to Ferrari and anyone else who was well-equipped. Mr Lee asked my advice and I told him that while we could increase the capacity to the full 2½-litres it could only be done by a complete redesign of the crankcase and many other details, which were beginning to show signs of trouble. I said that I could not guarantee a competitive power output, nor the reliability of what would now be a distinctly 'long stroke' variant of the original. But Mr Lee's enthusiasm gave him the urge to win once again, and between December 1st, 1958 and early March 1959, when the first engine ran, our design team made a remarkably fine job of the conversion.

This final stretch – or so we thought at the time – was actually a much more satisfactory job than the one it replaced, but we were still cautious and left the original cylinder-head and breathing arrangements as they were in order to put a natural limit on the power and to see how reliable it would be. I have always preferred to move ahead in small stages! Quite soon we *were* convinced, and the Mark II cylinder-head was produced with bigger valves and revised porting more appropriate to a full 2½-litre engine. This was the unit which led us to a lot of victories, and gave us all a great deal of satisfaction at Coventry. I believe that in 1960, its last full year in Grands Prix, an FPF never lost a race it contested. The interesting fact is that we produced no engine with more than 240 bhp (in other words only 96 bhp per litre) which was well under the figures we thought were only just competitive way back in 1954! This time, however, we weren't interested in the competition's claims – it was race results that counted.

I must be absolutely straight and say that almost all the big wins were to Cooper's credit. There were apparently several good reasons why this should be so, but one was certainly that John Cooper's engines were always installed strictly according to our recommendations, while Colin Chapman's were not! To be more specific – the mid-engined Coopers had our engines installed in a conventional layout (18 degrees canted to the offset to give carburettor clearance) while the first front-engined Lotus 16s, the 'mini-Vanwalls', had theirs canted at a very extreme 60 degrees. Not only did this mean that the engines were suffering all sorts of lubrication and centrifugal effects we had never anticipated, but the big Weber carburettors had to be remotely mounted on the chassis frame and connected by flexible piping to the original inlet manifolds. I feel it is only fair to say that we were not consulted about this at first, and that if we had been we would not have approved. There was only a single good thing to be said about the installation – it kept the frontal area down! Colin had problem after problem with these cars at first, and used to grumble at me about the low power output; I could only point out that he wasn't doing things the way we recommended, and when I asked him how it could be that Cooper were winning World Championship events with identical engines he couldn't really give us an answer!

Apart from all the Grand Prix activity, we seemed to produce all manner of other variants. The original 1½-litre and 2-litre units were still selling well, and Geoff Densham didn't seem to find any difficulty in getting rid of as many as our tool room could build. This, of course, was the irony of it all. Although the FPF has been called the world's first 'production-line' racing engine, there was precious little tooling to help us. Even though we built something like 273 FPF engines of one size or another, the only proper tooling aids were a few

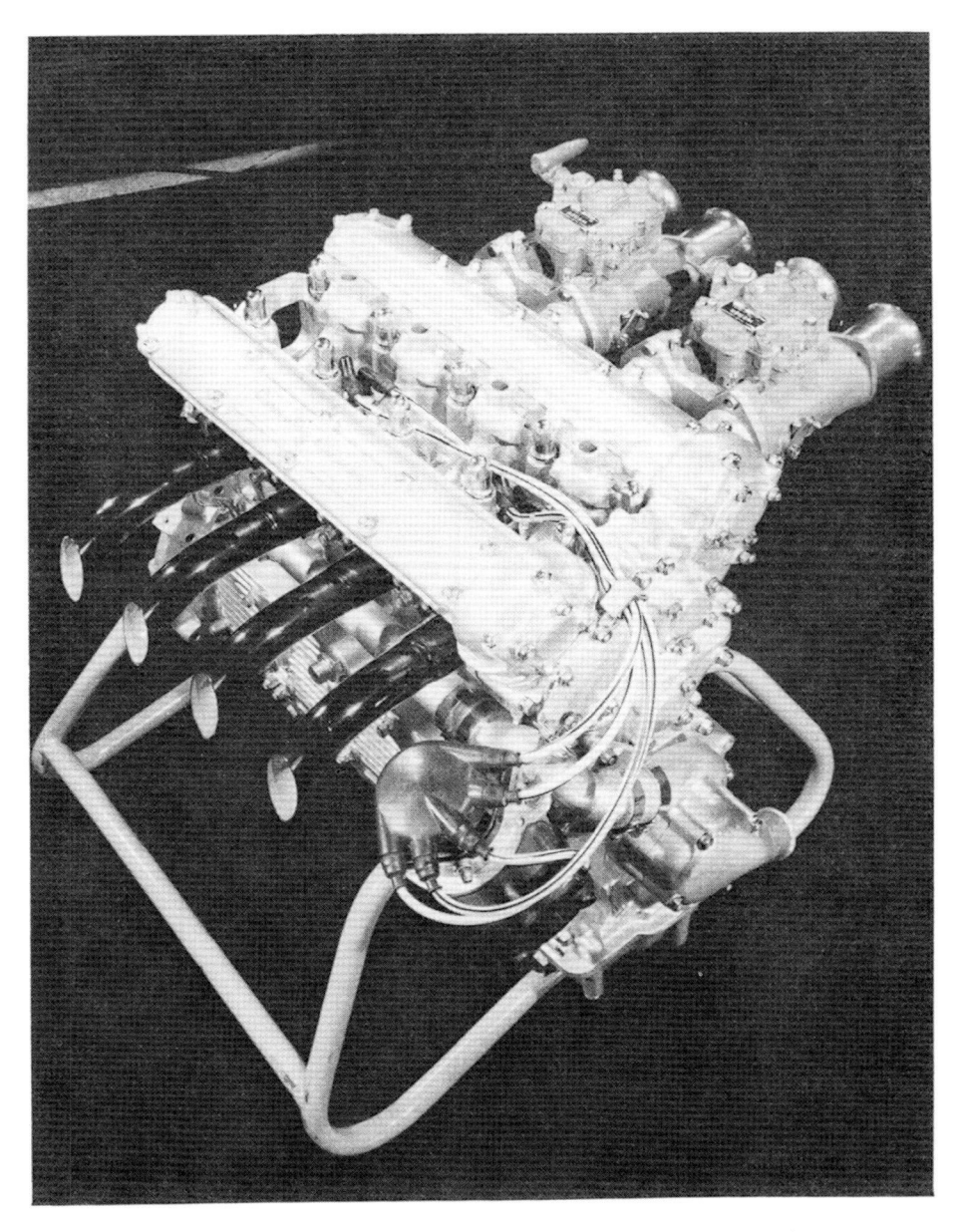

The legendary Coventry Climax FPF engine, which won so many races in so many different sizes. This is a 2½-litre example, which was show-prepared for Earls Court

It was a night for celebration when Coventry Climax were awarded the Ferodo Trophy for 1959. Leonard Lee, a partly hidden Graham Hill, myself, John Cocper, Tony Brooks and Les Leston surround the World Championship-winning FPF engine

drill jigs for machining cylinder-blocks and heads. Quite a number of components were made by outside suppliers, but every engine was carefully and individually built by one fitter, and all were power tested before delivery. All this placed a tremendous load on our limited resources at Widdrington Road, and I believe that even Mr Lee began to realise that his hobby and his patriotism were beginning to catch up with him. The 'small-scale' activity which produced so much wonderful publicity was now a very important part of Coventry Climax, and while I was as interested as anyone in keeping up with the constantly changing racing scene, there were many occasions when normal engineering work had to be side-lined in favour of a motor racing panic job.

Also, we were not making any money from our racing, indeed, at no time did the charges we made for engines do more than cover the cost of components; there was no way in which they could begin to pay for the design and development time which was expended. The constructors were continually pleading poverty, that motor racing was much too expensive, that they couldn't afford to go on much longer, and that their engines would have to be cheap. I could never really understand how they could go on suffering so much, especially when a few of them started buying private planes, but we did our bit. The first 1½-litre FPFs cost just £1,000, while the very last 2½-litre Grand Prix engines cost only £2,250 - a figure which wouldn't buy even a good Formula Three engine ten years later!

I said earlier that we all thought that the 2½-litre engine was the final stretch possible, but very much against my better judgment I was proved wrong yet again. Following their very successful testing at Indianapolis in 1960, Cooper decided to enter a car for the 1961 500-Miles race. Would we, they asked, be prepared to push out the limits of the old FPF yet again, to its absolute limits? I was also very much aware of the dangerous game of politics which was being played by the British constructors against the 1½-litre formula due to start on January 1st, 1961, and for which we had done no serious design work until then. The British wanted their Intercontinental Formula for 3-litre cars, even though there were no British engines anyway. No-one had asked us about the possibility of a 3-litre Coventry Climax engine, but I think certain people were hoping that I could wave a wand yet again.

None of us in the company wanted anything to do with the proposed Intercontinental Formula, especially as Mr Lee was sure the constructors would have to fall in line with the 1½-litre decision sooner or later, and would then want a brand new design from us. As a very special favour to Cooper and to Jack Brabham (who had won two World Championships with the aid of our engines) we agreed to make a special 2.7-litre FPF for Indianapolis. I was very anxious that the press and the constructors should not get the wrong idea and I wanted it clear that this was just a one-off. Accordingly, with Mr Lee's approval, I took the rather drastic step of issuing a statement to the press, of which the following were extracts:

"The advent of the new Intercontinental Formula, together with the publicity given to the Cooper/Brabham trials held at Indianapolis recently, has given rise to much speculation and rumour as to the possible modification of our 1960 2½-litre engine to allow for an increase in capacity to approximately 3-litres... The small dimensional increase which is physically possible leaves the capacity still well below 3-litres. It imposes considerably greater inertia loadings, and because of this the permissible crankshaft speed is reduced, and also, therefore, the potential power.... Following the trials carried out by Jack Brabham at Indianapolis John Cooper has persuaded us, and we have agreed somewhat reluctantly, to produce one special engine for this race in which the bore and stroke have been increased to give a capacity of 2,750 cc. Whilst this may prove an interesting experiment, we are not at all happy as to the reliability of the engine...."

In the same statement I also made it quite clear that we could not possibly afford the time to support both the Intercontinental Formula and the new Formula One.

However, these brave words were not enough. We finally bowed to the demand for other 2.7-litre engines, and after Brabham's splendid showing at Indianapolis in 1961 we made a few more. The engine was a bit more durable at Indianapolis than I had expected, probably due to the effect of the alcohol fuels they use there. It is well known that alcohol brews are much kinder to an engine, because the pressure rise in combustion is much slower than with petrol. But with very thin plated steel liners, and precious little water between them, we were really scraping the barrel for that extra bit of capacity.

There were many other happenings in this period quite unconnected with engine design. I certainly mustn't give the idea that I was only involved in FPF racing engines between 1955 and 1960. Far from it, in fact!

At the start of this period I moved the family out to a nice country house in Norton Lindsay, not far from Warwick (and incidentally very near to Bill Heynes, my old boss from Jaguar days), but by 1959 we were all back in Leamington, to the house where I am now writing this book, and in which I have now lived longer than anywhere else.

Bread-and-butter design work at Coventry Climax continued to be varied and interesting. By 1960 pressure from the Fork Lift Sales Director and from other members of the board was directed to the necessity of splitting fork-lift truck work from that of the engine and fire-pump offices, and it was decided to make Eric Buxton Chief Engineer – Fork Trucks. This relieved me of my immediate responsibilities on this side of the work. Buxton had been with Mr Lee for a number of years (coming to Coventry Climax from the Claudel-Hobson carburettor company in Wolverhampton) and had been involved in fire-pump design for some time.

As far as other work was concerned, I believe we set up some sort of a record in the industry in the 1950s. We probably designed more engines than any firm had tackled in a comparable period, most of which were technically successful and a goodly proportion of which went into production. There was no doubt that Mr Lee liked to have new engines around him. They were his great interest in life, but if he had been making millions of the same old design for too long I think he would have got very bored! Not that we designed engines just to keep him happy! We did a complete range of high-speed ohc diesels – three, four and six-cylinder in line, twin-cylinder units with counter-rotating bob weights to improve the balance, and two-stroke marine engines. Then, of course, there is the rugged little H30 two-stroke multi-fuel unit which we designed for the Ministry of Defence, and which now supplies ancillary power to every Chieftain tank made. This may sound as if I was never at home, but in fact I had gathered around me such a good team of designers that there wasn't much need for the burning of midnight oil. I got away for weekends fairly often, and I never took my work home. I found that as I got older I was very good at delegation, or giving other people enough work to do! But we could never have done any of this if we had lacked backing. Mr Lee was always behind us, and although he was no engineer – he never pretended to be – he had great faith in us and allowed us to develop our designs in the way I thought best. I could not have asked for a more pleasant arrangement.

Earlier I mentioned that originally I wanted to be a marine engineer, which was an ambition not realised. However, I loved 'messing about in boats', particularly if I could motorise one and use it for pleasure. Apart from my speedboat with the Gregoire engine in the 1920s (the hull of which had been used to set a cross-channel record a bit earlier) I didn't own another boat until I was well established at Coventry Climax. There, to my joy, I discovered that marine engines were being made, and that there was a 26-ft cruiser, with a Climax diesel engine, moored in the Solent but practically never used. Fortunately, both Mr Lee and I were very attracted to south-west Wales, and it didn't need much talking on my part to persuade him to have this craft moved down to Pembroke Dock where we could both enjoy it. Later, Mr Lee very kindly gave it to me, and it is now moored down at Falmouth where I have another house.

In the meantime, we also bought nothing less than the ex-Barmouth lifeboat! It only cost us £750, but we then spent a lot of time and effort in converting it into a really luxurious 32-ft cruiser. As we bought it, it was equipped with special RNLA horizontally-opposed petrol engines (built specially by the Farnborough Engineering Company, together with the rather inefficient and prehistoric jet propulsion screw shrouds for operation direct off the beach), but we replaced these by twin Coventry Climax diesels, which produced a really nice boat which would sail and could also be motored.

Later, when we got our racing engines going, there was an interest from the racing boat-building people who wanted to install them in their lightweight craft. Apart from the conventional boat-building concerns, we also had a very close relationship with the Healey family, who at one time in the 1950s were involved in building ski-boats. These were normally fitted with Austin engines and an American hydraulic gearbox. One year, however, Donald

Healey decided to enter one of his craft for the annual six-hour sports boat race, held on the Seine in Paris, and to make sure he had real performance he persuaded us to lend him an FPF racing engine! I suppose this engine (a 2-litre, producing about 180 bhp) was too powerful for the hull, which was only 15-ft long, but it made a very exciting craft of it, especially as Donald invited Tommy Wisdom to be the helmsman. Everything was going very well, the craft really going great guns, when suddenly it hit a half-sunken log, the hull was holed, and before poor Tommy really knew what was happening he found the water up around his chin and the boat going under. He thought he had better get out, and did so, but he first switched the engine off before it was filled with Seine water, and then made for the bank. The edges were so steep that he had to be helped out by kids who hauled him in with a branch.

When he got back to us he was dressed only in underpants and a blanket, so we thought we had to save his life in the traditional manner. Seine water was pretty foul and cold, so we quickly found a bottle of Scotch and set about decanting it into Tommy, who soon began to look a lot happier about the whole thing. That evening we all went out on the town to drown our sorrows. In Tommy's case he had been advised to drink a lot of milk, so our rule was that he had to alternate a glass of that with a glass of Scotch. Tommy suffered no ill effects, but the next problem was to find the boat, which was apparently sunk without trace. However, Tommy had had the good sense to mark the spot by landmarks, and after a lot of diving by a real diver complete with brass helmet and all the kit, it was finally located and raised in safety. It was such a sad day for Donald Healey and ourselves because we had all had such high hopes of winning the race, but we had found such a hilarious way of ensuring that Tommy stayed healthy that I almost felt ashamed of pointing out to the Healey family that the 2-litre FPF engine they had given a ducking was actually a new one, and had already been sold to a poor unsuspecting overseas customer! We had planned to run it in the boat as a good means of running in and test-proving it for our customer. But the engine, once recovered, was stripped and reassembled without further problems, and in due course a perfect racing unit was delivered!

There was one surge of interest in the FPF engine from British motor racing teams, but its usage really belongs to the next chapter. The CSI, in its wisdom, decided that they simply had to cut down speeds in Grand Prix racing (how many times some of us had heard that before!) and set out to do this by insisting on smaller engines and the use of pump fuel. This decision was greeted with howls of derision and abuse from the constructors, and we were not at all in favour of any change. But at least any engine we designed could be purpose-built for the formula, and would not continue to be a 'redesigned redesign' like the FPF, which nevertheless had done such a very workmanlike job in its time. For better or worse Peter Windsor-Smith and I had a new challenge on our hands, and could use a clean sheet of paper.

Coventry Climax FPE V-8 engine

Although the legendary Type FWA 'fire-pump' single-overhead-camshaft engine was the first from Coventry Climax to find its way into cars and to win races, the first racing unit to come from the inspired drawing boards of Harry Mundy and Walter Hassan was the FPE 2½-litre V-8, which was designed and built between the winter of 1952 and the summer of 1953. It was originally intended for sale to firms like Cooper, Connaught, HWM and Kieft, but as Walter Hassan has already explained none were delivered because its designers thought (incorrectly) that they had been unable to produce competitive power.

The FPE was a racing V-8 in the classical style apart from the relatively minor aberration of having complex hairpin valve springs instead of the normal coil variety. The five-main-bearing crankshaft used bolt-on balance weight inserts to six of the crank webs (something used later with great success on the larger-capacity FPFs when space limitations were becoming serious), and the main bearing caps were bolted vertically and horizontally to ensure adequate stiffness. Bore and stroke were 76.2 mm and 67.94 mm, respectively, for a swept volume of 2,477 cc, and there was provision for both bore and stroke to be increased when necessary, although neither Hassan nor Mundy had any thought for this at the time. The engine was, of course, designed specifically for the 2½-litre Formula One, but on the evidence of the last four-cylinder FPF engines there would have been no difficulty in stretching the FPE to at least 4-litres or even 5-litres if the scantlings had withstood the extra stresses imposed. The pistons and connecting-rods were similar to those used subsequently in the FPFs, two rods being positioned side-by-side on each of the four crankshaft journals. The original connecting-rods were split at 45 degrees to their main axis to facilitate withdrawal for rebuilds, but when it was discovered that the rather angular longer end of the boss was flexing under load with sometimes dire effect on rigidity (this was revealed later on the FPF version) the withdrawal facility was dropped in favour of a symmetrically-machined rod, which gave no further trouble. The pistons had a simple 'pent-roof' shape so that an 11-to-1 compression ratio could be achieved. There were two valves per cylinder, set at an almost symmetrically disposed 66-degrees included angle, the exhaust valves being hollow and sodium-cooled. Valve operation was by twin overhead camshafts per bank operated by the time-honoured bucket tappets allied to individually ground thimbles to give the correct clearance. Most people would call this the 'Henry' method, though Harry Mundy insists that Ballot should get the credit for this invention. Naturally there was a dry-sump system, and ignition of this none-too-high-speed design was by a Lucas racing magneto.

At first, carburation was to be via a quartet of twin-choke Solex instruments, but it took little testing to convince Coventry Climax that there was air-flow interference, and that the smoother-flowing properties of a fuel-injection system, however crude, should be adopted. All the significant power figures on this engine were achieved with SU fuel-injection systems, though it is interesting to remind ourselves that the fabulously successful FPF four-cylinder engines mostly used a pair of twin-choke Weber instruments, or sometimes a pair of the rare though generally successful twin-choke SUs. The firing-order on the first engine was set to give the effect of two four-cylinder units firing consecutively (at intervals of only 90 degrees) but this was found to give an impossible exhaust-pulsing condition, with much back-pressurisation and loss of power. A change to the more usual V-8 sequence was then made with immediate and startling improvements.

The engine ran for the first time at the end of the summer in 1953, and work continued into 1954, by which time Hassan and Mundy had become very depressed by the talk of 270 and 280 bhp from Formula One engines produced in Italy and Germany. If only they had known that the first $2\frac{1}{2}$-litre Maserati 250F units produced 230 bhp at 7,200 rpm, with maximum bmep of 179 lb/in^2 at no less than 5,800 rpm (Maserati never beat the later V-8 Climax figures, even in 1957 when using nitro-additives), and that the M196 Mercedes unit produced 256 bhp at 8,500 rpm in 1954 with 186 lb/in^2 bmep at a monstrous 6,300 rpm, they might have been more encouraged. There were troubles with the FPE, but these were mainly confined to the initial mistake in the firing order and in the poor performance of the hairpin valve-spring set-up.

Using an aviation petrol of 115/145 PN rating more than 250 bhp was achieved, and as soon as a 65 per cent mixture of alcohol fuel was tried with fuel injection a peak power output of 264 bhp at 7,900 rpm was obtained quite early in the development programme. This was significantly better than any engine built then, or until 1955, and was only exceeded notably before 1958 by Leo Kuzmicki's Norton-inspired Vanwall engine, for which 290 bhp at 7,500 rpm was regularly claimed. The other very significant features of the FPE included a dry weight of a mere 340 lb, and a torque curve which was substantially flat from 5,000 to 7,500 rpm, where the peak bmep was at least 190 lb/in^2.

Not to use this engine in a Grand Prix was, as it transpired, a tragedy, for it must surely have provided Cooper, Connaught and possibly Kieft and HWM with competitive Formula One cars. Hassan thinks that without the development of the four-cylinder FPF he would probably have pushed on with the FPE, which was at a very early stage in its development life. However, once the FPF project had got under way, and it became clear that sales would be large (by racing car standards) there was never time to go back to the FPE again. No further development was carried out, and the engines, spares, and unassembled parts were pushed into a corner of the racing department.

Quite unexpectedly, in 1966, the whole lot was sold off to Paul Emery, who had plans for converting the old engines to 3-litres (the engines had generously-dimensioned wet liners, easily convertible to larger bores) and fitting them to a Formula One car to meet the new 3-litre regulations. There was no support for this from the factory, who in any case had withdrawn from racing by then, and since the engines had not previously been installed in a chassis no-one gave Emery much chance of success. He proposed to add Tecalemit-Jackson fuel-injection (a simple continuous-flow layout then enjoying some success in racing cars) and offer the engines for sale. Apparently there were four complete units and many spare parts. Clearly he was not worried by the fact that the FPE had never been run on pump petrol, or that the compression ratio, valve timing and cooling arrangements would all have to be re-developed, for, at the British Grand Prix, in July, a strange car called a Shannon appeared for Trevor Taylor to drive, into the back of which one of these engines had been persuaded. In practice it was not a complete flop, for Taylor urged it around the 2.65-mile Brands Hatch circuit in 1 min 41.6 sec (7.5 seconds slower than pole-position Jack Brabham in his Repco Brabham, which was producing about 320 bhp at that time), but in the race the car failed to complete one lap. It was said to have retired with a split fuel tank. The old FPE engine never appeared again in public, and the Shannon went back into limbo. Hassan is glad about this, as he would not have liked to see any Coventry Climax engine performing without success in motor racing.

But an FPE engine is still to be seen in public – a show-prepared version in a glass case, which now lives in the Godiva public house, in Coventry. There would be no great incentive to steal it, as it is an 'empty' unit, and many of the visible components are mockups, although visually it is fully representative of the design.

There was a rival in Coventry at the time of the FPE's development, which is worth mentioning to clear up any misunderstandings. Councillor Harry Weston got together with racing-driver Leslie Brooke to produce a V-8, and one example was actually shown to the press. However, as far as is known this engine had very little running, never produced competitive power, and was certainly never raced nor fitted into a car. Nor, for that matter, was it ever connected with Coventry Climax's efforts, and neither Hassan nor Mundy ever saw the details of the design or the engine itself.

Coventry Climax FPF twin-cam 'four'

Hassan and Harry Mundy designed the famous Coventry Climax FPF engine in response to the new Formula Two regulations which were due to take effect in 1957 and specified unsupercharged 1½-litre engines which had to run on pump fuel, alcohol mixtures being banned. The basic layout of this engine was Harry Mundy's last task before he left the company to join The Autocar as Technical Editor in 1955. The company intended to supply all engines to a standard tune, so there would be no question of special units for favoured customers. In all respects the FPF was the first 'production-line' racing engine ever built in Great Britain.

We have already seen how the FPF developed virtually as one half of the original V-8 FPE designed in 1952/53. The cylinder-head assembly was unchanged apart from having the latest in helical valve springs, the connecting-rods were straight from the V-8, and so were the pistons and much of the valve gear. There was no question of using a half of the FPE's crankcase (people talk much too glibly of this process), and the four-cylinder unit had a completely new block/crankcase assembly, lubrication layout and accessory mounting positions. Magneto ignition was discarded in favour of a conventional coil and distributor, and there were dynamo and starter fitments when requested. Carburation was initially by double-choke SU instruments, though 40DCOE Webers were also under development, and were eventually preferred by most customers.

Bore and stroke were 81.2 mm and 71.1 mm, respectively, and swept volume 1,475 cc; this compares with 76.2 × 67.94 mm for the first FPE V-8. Consideration of gasket life and sufficient land between cylinders determined the small bore increase, which made a minor stroke change necessary to return to the correct $1\frac{1}{2}$-litre limit. Both Hassan and Mundy were fully aware that they had built in no reserves of space and strength to allow the engine to be enlarged at a later date. Indeed, the original stretch, to 1,960 cc for the Jack Brabham Cooper-Climax at the 1957 Monaco GP, when it held third place until a component failure a few laps from the end, was made reluctantly. Hassan said then that as far as he was concerned the absolute limit had been reached! Assembly of that first engine was completed by Alf Francis at the factory, and the engine then received just two days of running, setting-up and power-testing, followed by a rush back to Surbiton for fitment to the race car.

Almost at once, therefore, there were two versions of the FPF. The $1\frac{1}{2}$-litre F2 unit was in production for many customers, producing a guaranteed 141 bhp at 7,300 rpm, and a competent peak bmep of 182 lb/in^2 at no less than 6,500 rpm. By Hassan standards that was a high figure (he has since heaped scorn on designers who cannot produce racing engines with good mid-range torque) but the shape of the torque curve told a typically-Coventry Climax story. Bmep was above 170 lb/in^2 between 4,000 rpm and 7,500 rpm, and above 180 lb/in^2 from 4,500 to 6,800 rpm — nothing peaky or undrivable about that. The specific output of 96 bhp/litre was unremarkable, except that it was achieved without recourse to alcohol fuels. The engine weighed only 280 lb.

The 1,960 cc unit was a rushed compromise, which meant there were no changes to the head, valves or porting, though the carburettors were re-set to cope with alcohol fuels (Formula One still allowed it). The new bore and stroke (86.4 mm × 83.8 mm) necessitated new pistons, liners and crankshaft, but changes were otherwise minimal. Peak power was a mere 176 bhp at 6,500 rpm, with an altogether excellent bmep figure of 204 lb/in^2 at 5,000 rpm — certainly a new peak for any unsupercharged engine. This achievement finally convinced Hassan of the worth of under-sized ports and higher gas velocities for producing lots of healthy drivable torque.

The FPF was instantly successful in Formula Two, particularly in Coopers.

Their only effective competition came from Ferrari, whose car was very powerful (some said 180 bhp) but front-engined and heavy. However, the factory's problems were just starting. If they had not been pressured into enlarging that first FPF to 2-litres the story might have ended here, but the ever-hopeful car constructors were not content with that. For 1958 Hassan and his Sales Manager, Geoff Densham, were persuaded into making a few even larger units. First of all a single 2,015 cc engine (like the 2-litre but with oversize pistons) was supplied to Rob Walker, and distinguished itself by winning the Monaco GP with Maurice Trintignant. This merely confirmed Stirling Moss' startling success in the 2-litre car in Argentina and was a delightful baptism of success for Coventry Climax, who were soon to get very used to this sort of thing.

However, Hassan's main weapon in 1958 was to be a quartet of 2.2-litre engines – two each for Cooper and Lotus. The extra capacity was squeezed out under some duress by the designers. Bore and stroke became 88.9 mm, giving a capacity of 2,207 cc, and the only way this could be accommodated was by recourse to a packing piece between the top of the block and the underside of the head, with the new cylinder liners a slip-fit into the lot! It was a brave 'sandwich' solution which could only work with regular and rigorous inspection. This version produced 194 bhp at 6,250 rpm on the newly-compulsory Avgas, with peak bmep of 204 lb/in^2 at 5,000 rpm.

No changes had yet been made to the basic block or heads, and accommodating long-throw crankshafts meant that balance-weight dimensions were limited. The 2.2-litres were very rough, and although strictly rev-limited there was early evidence of fatigue cracking around the centre bearing webs, which was only cured by the use of steel straps over the bearings. Thus equipped, the Coopers had a successful season, amassing 31 World Championship points, and a best placing of third in the German GP. The beautiful 'mini-Vanwall' Lotuses failed miserably because Colin Chapman insisted on the engines being installed at 60 degrees to the vertical (to keep the centre of gravity down); invariably they overheated, gave lubrication troubles, or had difficulties with the remote-mounted carburettors which this installation made essential. It was not until Lotus introduced their much simpler, mid-engined car for 1960 that they became a power in Formula One racing.

Meanwhile, Hassan had taken a big decision in 1958. It was known that the Vanwalls would probably retire at the end of the year, and both Cooper and Lotus were anxious to tackle the Ferraris on equal terms with full 2½-litre engines. Leonard Lee was full of enthusiasm for the project and sanctioned the complete redesign that was necessary. There was to be a new cylinder block, with bore and stroke of 94.0 mm and 89.9 mm and a capacity of 2,495 cc, the dimensions being dictated by available space rather than any erudite theoretical considerations. Peter Windsor-Smith also had to find space for adequate crank balance-weights, solving the problem by specifying an expensive GEC heavy-alloy material which was bolted to the webs. The revised head was at first 'under-valved', to hold down output in case the engine's strength proved to be lacking, but early experience soon *allowed revised heads with larger ports and valves to be used. Peak power was a very creditable 240 bhp at 6,750 rpm, and peak bmep was up to 210 lb/in^2 - once again a new record for unsupercharged engines.*

In spite of the 2½-litre's worldwide success, it was not the subject of lengthy deliberations. Detail design began on December 1st, 1958 and the first unit won a race at the Easter Goodwood meeting on March 30th, 1959 (in Stirling Moss' Rob Walker Cooper). For the next two years there was little to match the works Coopers of Jack Brabham and Bruce McLaren and the Rob Walker Cooper and Lotus driven by Stirling Moss. BRM and Ferrari might have had more power, but it was neither reliably delivered, nor were their torque curves of such desirable shape, while their cars were heavier and mainly front-engined. In 1959 and 1960 the 2½-litre FPF powered 13 winning cars and many runners-up in World Championship races.

The 2½-litre formula was wound up at the end of 1960, to be replaced by the very unpopular (at that time) 1½-litre formula announced by the CSI. The FPF's career would have been at an end if it had not been for the incredibly head-in-the-sand attitude of British constructors, and for Jack Brabham's first excursion to Indianapolis. British car makers opposed the new formula so vehemently that they proposed a complete boycott and the introduction of a new 3-litre Intercontinental Formula which they hoped would attract the Americans. While Ferrari got on with developing new cars and his well-proven V-6 engines to take the World Championship in 1961, Cooper, Lotus, BRM and others hoped their voices would kill off the new formula.

They failed, and when this became clear they rushed to Coventry Climax for help. Climax's long-term answer - the FWMV - is discussed in a future chapter, but as a stop-gap for 1961 it was agreed to supply a limited number of up-dated FPFs in 1½-litre form. The Mark II used all the best constructional features of the 1960 engines, but with the 1½-litre dimensions, apart from a slightly increased bore (to 81.8 mm). Better breathing and revised valve gear allowed a peak speed of 8,200 rpm - the best ever seen on an FPF. There was no hope of this interim engine matching the V-6 Ferraris, or the V-8s that several firms (including Coventry Climax themselves) were known to be developing, but for the British companies it was a case of Hobson's choice. Everyone including BRM used the Mark II, which boasted 152 bhp at 7,500 rpm. In spite of a huge power deficiency against the Ferrari - put at 20 to 30 bhp - Stirling Moss drove two magnificent races at Monaco and the Nurburgring to win for Rob Walker in his Lotus, while Innes Ireland gave Team Lotus their first Grand Prix victory in America at the end of the year, when Ferrari were absent. It was also the last Grand Prix to be won by a four-cylinder car; for 1962 the V-8 FWMV took over.

Cooper's visit to Indy started with a casual testing trip after the 1960 United States GP, when Brabham's 2½-litre car lapped at over 130 mph. The sequel to this was that Cooper pleaded for yet another FPF-stretch so that Brabham could race at Indianapolis in 1961. As will be obvious from previous comment it was an extremely difficult task, and it was only by audacious risk-taking and selective assembly that a couple of very special 2.7-litre units (with increased bore and stroke) were made available. The company's press statement, designed to discourage further requests for a similar engine, failed in this respect, and before the year end two more were built, while the count finally totalled 14 engines in addition to a few old 2½-litre engines which were converted to 2.7s. Jack Brabham's raid on Indy was successful, as his diminutive Cooper finished a strong ninth against the 4.2-litre monsters.

Future motor racing historians may not revere the FPF's memory because it broke no new ground technically. However, its own success story is written in numbers built. There were 273 engines of nine differing capacities between 1957 and 1965. No fewer than 159 of these were 1½-litre units, and 48 were full-sized 2½-litre Grand Prix engines, a Formula One production run not exceeded until the Cosworth Ford DFV juggernaut rolled into action in the late 1960s. One can think of no other unit which won so many races in so many different guises, unless it was Lampredi's V-12 Ferrari. But popularity brought problems, and future Coventry Climax participation would have to be on a more selective basis. For his new engine – the V-8 1½-litre FWMV – Hassan was determined to limit his customers, and he wanted the factory to keep an eye on every one at rebuild time.

7
The climax for Climax

There is one date in motor racing history that some of us will not forget in a hurry. That date was October 1958, and the place was a social occasion at the RAC offices in Pall Mall. The happy and predictable part of the evening were the presentations made to Tony Vandervell and Mike Hawthorn for their successes during the year, but it was also known that the CSI would announce changes in the Grand Prix formula.

The President of the CSI, Monsieur Perouse, could not speak English, so the announcement had to be made on his behalf by Pat Gregory of the RAC. To the complete consternation of all present, he stated that as from January 1st, 1961 the Formula One engine size would be cut to 1,500 cc – in other words to the size of the Formula Two cars for the past two years. The reaction was explosive and the meeting disintegrated into uproar. I can do no better than quote one of the most experienced motor racing writers, who said: "This announcement was greeted by a complete hush – then cries of 'shame' and unrestrained, prolonged booing. The general feeling was that the decision was aimed largely at putting an end to the present British supremacy – and that a step had been taken that could well result in the demise of Grand Prix racing".

Certainly we knew that there were to be changes, because several of us had been consulted during the summer as to our views, but we never expected anything like this. I had been invited to several meetings at the RAC and at the CSI in Paris to state Coventry Climax's preference, which was really very flattering as at that time we had still not designed the full 2½-litre FPF engine and no full-sized Coventry Climax engine had appeared in a Grand Prix car, although Rob Walker's Coopers had already won the Argentine and Monaco races with 2-litre versions.

At Coventry Climax we were not in favour of any changes in the engine limit. We did not see why the whole basis of Grand Prix design had to be changed on a particular date, and we were all for continuing the existing 2½-litre limit. On many occasions we had had it drummed into our memories by constructors such as Colin Chapman and John Cooper, not to mention the journalists, that the cost of racing was very high and increasing all the time.

Engines were stated to be the most expensive item, which was reasonable, although we always tried to supply our products at the minimum possible price. I knew that if the Grand Prix formula was changed, either by allowing bigger engines or by imposing a lower limit, then the company would be faced with the need for a completely new design, and I also knew that this would have to be capable of a higher specific power output. It went without saying that a higher specific output meant a more complex engine and higher costs, something our customers would be reluctant to accept.

Therefore, whenever I was asked for my opinion about a new formula, I always opted for retention of existing limits, as I thought there was plenty of scope for development of the 2½-litre cars. I certainly did not agree with change just for the sake of change, which I suspect is what a few of the constructors were keen to see. As for the circuit owners and promoters, they thought the existing formula had gone on for too long (it was then in its fifth year) and they felt that a change was needed to inject novelty and bring about an increase in their attendances. The important difference between owners and constructors on the one side, and the CSI on the other, is that those actually involved in the racing always wanted to see *bigger* engines, which would make the cars more exciting.

If the British constructors had accepted the new 1½-litre limit with good grace there would have been time for Coventry Climax to develop a new design into a state of competitiveness by the start of the revised formula, but for the next couple of years Cooper, Lotus and BRM (who were not our customers, of course), sometimes helped by Tony Vandervell, continued to lobby for the 1½-litre formula to be shelved, and threatened a boycott which they thought would kill it before it got off the ground. After all, they reasoned, Grand Prix racing in 1952 had been run for a different formula because there was no enthusiasm for the existing Formula One, so why not in 1961? They produced proposals for an Intercontinental Formula, which would be for 3-litre cars in the hope that it might attract entries from America, but although a few races were run to this formula it did not prove popular with circuit owners. The interesting thing is that none of the constructors who wanted to run cars to the Intercontinental Formula ever bothered to inquire if Coventry Climax could, or would be willing to, supply suitable engines. When we made it clear that we could not support them, and that there was absolutely no way of expanding the FPF engine to 3-litres, they were rather annoyed!

Meanwhile, the CSI resisted criticism stubbornly, and the only change they made to their 1958 proposals was a reduction in the minimum weight. I was convinced that we would have to design a new engine for a new formula, but we obviously could not afford to begin design and development work while the regulations were still under discussion. When at last the die was cast we had lost some 18 months of valuable time and there was no longer any possibility of having a brand new 1½-litre engine raceworthy by the beginning of 1961.

Of course, we had the 1½-litre four-cylinder FPF engine, but Peter

Windsor-Smith and I were convinced that it would not be good enough for 1961, even though the 2½-litre version had won a lot of races in the previous couple of seasons. Whatever I thought about the merits of good torque delivery and simple construction, there was every indication that Ferrari and BRM were planning multi-cylinder engines whose specific outputs were way ahead of anything we had achieved so far. In addition both had started to use lightweight mid-engined cars. I might have been satisfied with specific outputs of 100 bhp per litre in 1954, but for 1961 I thought that we would need something in excess of 120 bhp per litre. We could not see any way in which the FPF four-cylinder could be persuaded to give us this, so a new engine was essential.

Deciding on a new layout was simple. The pundits, journalists and enthusiasts who would not have to do the hard work were all saying that complex little multi-cylinder designs would be essential to ensure high enough power outputs; we agreed with them up to a point, but not to the extent canvassed. People were talking confidently about 12-cylinder or even 16-cylinder units, but we were sure that an 'eight' would be adequate and could be more easily developed; we reached the same conclusions as for the 1952/53 FPE. It was both interesting and gratifying to learn that BRM had come to the same conclusion, and that their engine would be similar in many ways to ours and would produce very similar results.

In the few years that Coventry Climax had been involved in racing the cars had become much smaller and lighter, and for the new formula, where a small frontal area was essential, we knew that our engine should also be as small as we could make it. We were very lucky in having the bare bones of a suitable engine, which had evolved over the years from a tiny little fire-pump layout made to succeed the original featherweight of 1950/51.

As I have already recounted, the featherweight FW engine was designed speedily and successfully by Harry Mundy and myself, and while we were happy with its performance we knew that it had a substantial 'factor of ignorance' built in; in fact we were able to improve the original 35 bhp unit to more than 100 bhp for racing purposes. To succeed it, and by drawing on all the experience of the intervening years of racing and industrial production, we developed a much smaller engine, one of only 653 cc but with a single overhead camshaft. We also had another scheme which involved using this little engine as a marine outboard unit. In the course of development we concentrated on marine work, which meant that we arranged for the engine to be efficient when running vertically or horizontally, and as a project we dubbed it FWM, or 'featherweight marine'. During its work programme the capacity grew to 745 cc, but in spite of its great efficiency it proved to be too expensive for marine outboard sale in competition with the imported American two-strokes. The MEXE at Christchurch carried out many trials in boats up and down the Solent, but eventually the sheer size of the American marine engine industry defeated us. Our Government had decided to take away all import restrictions

Jack Brabham's part in putting Coventry Climax engines on the Grand Prix map should never be underestimated. Here he is using a 1½-litre FPF engine to power his Formula 2 Cooper, the car which was the Surbiton company's forerunner to their first Formula 1 car

Already a World Champion, and heading for his second successive title. Jack Brabham heading to victory in the 1960 Dutch Grand Prix, this time with his FPF engine in full 2½-litre form

The first version of the Coventry Climax FWMV V-8 1½-litre Formula 1 engine. The Weber carburettors were used in 1961 and 1962, after which they were replaced by Lucas fuel injection

on American outboards, and that rather cut us off. Weight-wise, power-wise and cost-wise we simply couldn't compete, because the tremendously high production achieved in the United States allowed everything to be die-cast. Fashion also entered the picture, because American manufacturers were able economically to introduce new and spectacular models every year.

However, the little FWM was a considerable success as a fire pump, not only because it was powerful and reliable but because it was very much lighter and smaller than our first effort: I believe we saved over 100 lb, and of course where only two men were to carry the complete pump such a saving was much appreciated. The units were stowed inside fire appliances in the same way as the first FWPs had been.

All this took time, but since we were now very well connected with the motor racing fraternity I suppose it was obvious that someone would see a motor racing application for it. Colin Chapman had already persuaded us to produce a 750 cc version of the FWA – the very short-stroke 'one-off' engine used for Le Mans in 1957 – but for the following year we made a racing version of the new mini-fire-pump, which we called the FWMA. The FWM was in many ways similar to the old FWA in that it had a single-overhead-camshaft layout and a wedge-head, but it was simplified by elimination of the jackshaft and was smaller. Right from the start it was clear that it could be competitive due to its light weight and smaller reciprocating parts.

At about this time, incidentally, the FWMA design was adopted and modified by Rootes, with our approval, for use in their Hillman Imp. As productionised by Rootes (incidentally under the direction of Leo Kuzmicki, who was closely involved in Vanwall racing engine design) the engine grew to 875 cc, and used die-cast cylinder-head and block, but was still very close to the original layout. In the Imp, which is still in production as I write, the engine is installed at an angle of 45 degrees. The original tie-up was via Peter Ware (then Technical Director of Rootes) and Mike Parkes, who was Development Engineer, and was already making quite a name for himself in Coventry Climax-powered sports-racing cars, as well as Ferraris!

In 1960, when we decided to go ahead with the new Grand Prix engine design, I visualised a small and simple V-8 based on the same sort of cylinder dimensions as those already existing in the FWMA. It seemed logical to prove out our plans for the cylinder-head and breathing arrangements on the small four-cylinder design, and as part of the development programme for the V-8 we designed and developed a twin-cam head, with hemispherical combustion chambers, as a conversion to the racing FWMA. This proved very encouraging almost immediately, for it produced 83 bhp at a rather modest 8,200 rpm – equivalent to a specific power output of 111 bhp per litre. Although this was not enough for the Grand Prix engine, it was definitely promising and it proved that our plans for the V-8 were basically sound. Lotus used examples of this little FWMC in an Elite at Le Mans, in 1961, but the car retired with an oil-pump failure. However, by this time we were nearly ready to run the V-8 as the FWMC had been running since the end of 1960.

Various people have suggested that the V-8 was really just a doubling-up of the FWMC, but this is not really true. Certainly we used the basic porting, valve gear, camshaft drive and cylinder-head details from the FWMC, but the bore and stroke were different, the cylinder-block construction was revised to allow open-deck casting, and most important, we used a five-bearing crankshaft instead of the FWMC's three-bearing crank, and stretched the engine to accommodate the extra big-ends. So the cylinder centres were not even the same as on the FWMC, and the cylinder-heads consequently were not interchangeable.

All this was going to take time, and it didn't look as though our V-8 would be ready to race until the end of the summer, so in the meantime all the British racing car constructors were screaming for engines they could use right from the start of the 1½-litre formula, for even BRM were in the same parlous state as their V-8 would not be ready until the end of 1961, either.

Mr Lee was still most enthusiastic in his support of motor racing, and when I suggested that one way to tide over the constructors for a season would be to supply them with updated 1½-litre FPFs he readily agreed. We were also surprised and rather flattered when we had a call from Raymond Mays at BRM asking if we would supply them with FPFs for the season while their own engine was being developed! I wonder if the situation would have been the same if the boot had been on the other foot? This meant that for the 1961 season, *every* British Grand Prix car would be using the same make of engine. Not that we could guarantee much success, because we knew that Ferrari would be using his Dino V-6 engines which were producing around 180 bhp; even the best of our FPF Mark IIs could only push out around 152 bhp, though this was allied, as usual, to a healthy torque curve. Those FPF Mark IIs combined the fully-developed constructional details found to be so satisfactory on the final 2½-litre engines, with virtually the same bore and stroke as first employed in 1957. They may not have been truly competitive with the Ferraris, but they pleased me on several counts, and were the only FPF variants ever to achieve more than the 100 bhp per litre I considered the minimum for respectability, and this was on commercial '5-star' pump fuel.

We only intended the FPF Mark II to serve for one season, and although I was quite sure the cars so fitted would be make-weights I was delighted when Stirling Moss drove his heart out to win at Monaco and at the Nurburgring, and again when Innes Ireland won the American Grand Prix at the end of the year. But those were the only three wins out of the eight events contested – the year before we had won every time we appeared.

During the winter of 1960 and spring of 1961 we pressed on with the new Grand Prix engine, although it had to take its place in the queue around our industrial work, the production and testing of more than 30 Mark II FPFs and the considerable testing required to make one rather special 2.7-litre FPF for Jack Brabham's Indianapolis car. We based as much as possible of the

FWMV's detail on the little 750 cc FWMC engine that we knew to be successful, but the complications of a 90-degree V-8 layout made many changes essential. We had to use a quartet of downdraught twin-choke Weber carburettors which were new to us, and because we opted for a conventional two-plane crankshaft we were faced with a ticklish problem connected with the tuning of exhaust-pipe lengths and interconnections to ensure that we benefited from the maximum extractor effects. Ironically, though, the first important problem we found when the V-8 engine was run came about because I broke one of my own rules, namely to alter a successful detail without sufficient proof of its successor's worth.

Coventry Climax, like other eminent racing companies, retained the services of a consultant who was a long-time expert on cylinder-head design and air-flow through ports. I will spare his blushes, because he has done much satisfactory work for us at many other times, because he is still active in business, and because I have rarely failed to remind him of this lapse on social occasions in recent years! We had completed our cylinder-head design, and as usual we had had a wooden model 'flowed'; in non-engineering terms this means that we had carried out tests to check how much air would pass through the ports under certain standard conditions. A week or so after this we received the results of satisfactory tests, together with a sketch of changes which our consultant claimed had improved the air-flow considerably. As his advice had usually worked out well in the past we incorporated the same shape and profile in the cylinder-head patterns. Eventually, the first prototype engine was built and put on test for the first time in May 1961, when we were disappointed to find that it was considerably down on expected power. We had a lot of trouble in settling on a satisfactory exhaust system, but it was some time before we were able to determine the reason for the poor performance. It wasn't until we sat down to check detail for detail against the first FWMC four-cylinder twin-cam head that we realised that perhaps the last-minute inlet port changes had made a difference. The simple and quick solution was to cobble-up a rather precious cylinder-head by boring out the original profile and inserting a steel tube in its place. The result was an immediate improvement, with more-or-less twice the power of the little four-cylinder engine, so we reverted to our original design at once.

Getting a satisfactory exhaust system took much longer. Because we had specified a conventional 90-degree crank - that is to say one with eight individual crankpins - to eliminate secondary vibrations, the exhaust impulses were rather oddly spaced. We tried the old FPE remedy of fitting individual stack pipes (as did BRM, who used them on their cars for a time in 1962) but this resulted in a satisfactory power output at the expense of a very peaky torque curve, which was not acceptable as it gave the engine a very narrow useful rev-band.

We tried all sorts of layout - two-into-two, four-into-two, four-into-one, all sorts of combinations - but it was not for some weeks that Gray Ross made the breakthrough. Running the engine on the test beds, he finally evolved the serpentine layout which appeared on the customers' cars at first in 1962 and 1963. This effectively connected the inner pair of cylinders from one bank with the outer pair from the other bank, and *vice versa*, the whole exhausting into twin pipes, high up above the transmission. In this way the exhaust pulses combined to produce higher power outputs *and* a better torque curve than any other system we tried. This layout was criticised by many and frankly derided by some, but no-one has yet been able to offer a better solution. BRM evolved a 'flat-crank' engine as soon as possible to eliminate their use of vertical stack pipes, but on the other hand I was rather gratified to note that Ford of Detroit adopted our 'cross-over' layout for their Indianapolis cars a few years later, mainly at the instigation of Colin Chapman, who was to run their cars.

By the end of the summer we were getting a very satisfactory performance out of the engine, although I must also admit that we were getting a very unsatisfactory number of complaints about the noise of testing from local residents in Coventry. On weekdays, things were not too bad during the day, but there was so much pressure on us to get this engine out that we had to do tests in the evenings and at weekends, when things were much quieter. We had our regular protesters, in fact we could almost predict who would complain, depending on the wind and the weather, but of course they had a point. I am sure that I would have complained if I had lived nearby! All I can say now is that the results of all the noise were satisfactory. From the first runs in May, when we were nowhere near the power I knew to be necessary, to the July holiday period we worked up to a figure of 174 bhp - 116 bhp per litre. This was still a little below my initial target, but in view of what I knew, or thought I knew, about the V-6 Ferraris, I decided we ought to start actual racing trials, which might even turn into victories at this early stage.

We didn't have engines to throw around, so our very first prototype unit had to be prepared to race, and while it was away we finished building the second one. Jack Brabham was the obvious choice to try the first engine - by then he had twice become World Champion driver in a Cooper, and was a very good tester-mechanic into the bargain - and by dint of some furious and concentrated work we managed to get it down to Surbiton in time to be built into Jack's Cooper for the German Grand Prix. There was very little time for development, so the engine was just installed as quickly and neatly as possible, and sent off to the Nurburgring.

However, it was a pretty disastrous start to the project. Not only did the distributor drive shear almost as soon as the car started in practice (*that* hadn't happened on the test bed), which meant that Peter and I got our hands dirty acting as mechanics and engine rebuilders, but we began to experience a mysterious water-loss problem which at first we put down to simple overheating. It was very easy to plump for overheating, as the engine was certainly boiling whenever we saw it stop at the pits, by which time a lot of water had disappeared, but no matter what John Cooper's mechanics did to the cooling

I don't seem to be very convinced by the yarn Dunlop's Dick Jeffrey is spinning, while Peter Windsor-Smith, on my right, looks non-committal

Celebrating another World Championship with Dunlop and members of the Press. John Blunsden, sandwiched between broadcasters Maxwell Boyd and Robin Richards, represented *Motoring News* that night, is now my publisher

system and the radiator they couldn't cure the problem.

We never finished a race with this engine in 1961, though Jack Brabham led the United States Grand Prix for more than half the race before the water-loss problem let him down again. I was very puzzled, especially as we took every precaution, including centrifuging the water, and taking even more care than usual over gasket-fitting and engine-build. We never had this trouble when the engines were on the test bed, and frankly it wasn't until we mocked up an actual racing car cooling system around the engine at Widdrington Road that we realised what was happening. It all stemmed from a decision of mine to design the cylinder-block so as to simplify the water jacket cores; we decided to use open-deck cylinder-block casting, and chose wet cylinder liners which were seated on a ledge at the bottom of the water jacket in the crankcase. At the top these liners were not supported radially, but were fitted with Cooper rings to ensure a seal to the gasket when clamped down by the cylinder-head, like on the Citroen and Standard Vanguard engines.

When we rigged up the car-type cooling system, along with some Perspex pipes, all went well until the engine reached a certain temperature, when suddenly a stream of bubbles appeared and the system began pressurising and losing water. There was no drop in power output, nor in reliability, but the water was being pushed out. In a car with a sealed cooling system this would be disastrous, because once the system pressurised and lost water overheating would ensue, followed by a seizure and a lot of damage. Once we stripped the test-bed engine and realised that the Cooper rings were shuffling and the cylinder-head was actually lifting slightly (the gasket was not blowing, but its clamping loads had gone) it didn't take us long to realise that the wet cast-iron liner was growing at a different rate from the aluminium cylinder-block as it warmed up, which allowed the seals to be weakened. Once we were sure we could have kicked ourselves, for with a test-bed cooling system in use we would never have seen the evidence. Test beds usually operate on a total-loss cooling system, with a tap refilling the cold water tank at the rate at which water was being passed through. In our case pressurisation was pushing combustion gases into the water, but of course it did little to raise the temperature, and could not pressurise an open water tank.

It all proves what I have often said in my papers to engineering bodies and to young engineers in their training, namely that one never finishes learning and should never become complacent.

Once we knew exactly what was happening it was a relatively simple matter to fit an aluminium sleeve outside the liner, which supported it and eliminated the differential expansion; water loss or overheating was never again a problem with healthy FWMV engines.

All this happened during the winter of 1961/62, after which we had to set about building engines for our customers. By August 1962 our racing staff had built up 16 engines for sale (which we identified as Mark IIs), while we hung on to the first two for a while for further development. By modern standards the cost of these engines was almost ridiculously cheap; they were priced at £3,000 each, which was still thought to be fairly steep at the time. I suppose the constructors had got used to taking their FPF engines off a 'production line' at £2,250 a time, and thought the much more complex FWMV should have cost them no more! There was, however, no question of Coventry Climax going into the racing-engine production-line business again. For one thing, we were still building and selling a fair number of FPF engines, though our Formula One activity with them had almost ceased except for the Tasman series, and we were still busily building automotive engines for Lotus and Jack Brabham (the featherweight FWEs). Coventry Climax Grand Prix engines were always looked after by Harry Spears' department at Widdrington Road, and this meant that there was a definite limit to the number of engines with which we were able to deal.

Motor racing teams were not operating on a shoe-string, by any means, but by comparison with today they never carried too many spares. Cooper and Lotus, for instance, with two-car Grand Prix teams, only had three each of our new V-8 engines in 1962, two being installed in the cars, perhaps with the spare in the transporter. This meant they could only afford one engine blow-up per event before they had to think of cannibalising one good engine out of two duds, or of scratching a car. There were a few blow-ups of our engines, it is true, which were due mainly to connecting-rod failures at first and over-exuberance in later years when we had made them fairly reliable. Connecting-rod failures often resulted in holed cylinder-blocks, but as I recollect we always managed to repair these by patching and welding, except in one absolutely disastrous instance. But it certainly wasn't meanness which meant that we repaired rather than replaced the blocks. Personally, I would have preferred to supply new castings rather than patch up the damaged ones, but it took so much time and cost to produce a tiny number of new cylinder-blocks and heads that we always managed to repair instead. Fortunately, the results bore us out, and the repaired engines always performed as well afterwards as they had originally. There was no easy way of disguising the repair, either – inspection of those fitted to cars now in museums would probably show up the evidence in one or two cases. Later, of course, we laid down several new engines for 1963 and 1964, but these were not meant to replace existing units.

We experienced one intriguing little phenomenon with the FWMV which, as far as I can remember, never cropped up with the earlier four-cylinder Grand Prix engines. When we had built a brand new V-8 and ran it, it would tend to stiffen up a little. It was not so much as to be noticeable on the test-bed, but when we started the strip down for inspection we found that we couldn't rotate the crankshaft by hand. It was Rolls-Royce, once again, who helped us out of the problem, as this sort of thing had happened to them. They forecast correctly that we would find that the crankcase casting would require some further heat-treatment over and above that normally considered necessary for stress-relieving. Rolls-Royce had developed this wrinkle for their own use some time earlier; they found that when complicated crankcases were built and machined, they were perfectly satisfactory initially and allowed the crankshaft to spin freely, but after a few hours' running there was enough distortion (almost as if the casting was slightly arching its back) to cause the tightening-up we had noticed. After the second stress-relieving treatment, we had to re-bore the main bearing line to complete the work – the discrepancy was only a matter of one or two 'thou', but if ignored it would certainly have caused a loss of power and bearing failure.

Almost from the start of 1962 this engine did very well, but in Grand Prix records it might appear that only Jim Clark had the best engines! We never consciously gave anyone, or any team, the best engines, for this would have made it very difficult to deal with other customers. Indeed, we had a minimum standard which had to be achieved by every engine we let out of the factory after a build or rebuild. However, from time to time we made improvements, and in this case we usually let Colin Chapman try one out at the same time as we supplied one to Jack Brabham. At that time Jim Clark was noticeably faster in his monocoque Lotuses than anyone else in Grand Prix racing, and since I always used to say that (like Stirling Moss) he could beat the rest even if his engine was slightly below standard, it is hardly surprising that he won most of the races with good engines! We only broke the rule of 'same for everybody' in our last year, but by then we were sure that Clark was well on the way to his second World Championship, and therefore really deserved our best attention.

The interesting point about Jim was that although he was a supreme driver – in my opinion the very best who has ever been wheeled out on to a race track, then or since – he was not sensitive enough in the engineering sense to tell us much about the behaviour of our engines. He built up a remarkable relationship with Colin Chapman, and could describe very precisely his thoughts on the drivability of the car itself, but he seemed unable to tell us much about the operational characteristics of our engines. This contrasted very strongly with Jack Brabham, who as I have already mentioned was an excellent mechanic as well as a world-class driver, and could be relied on to give us a full report. Generally speaking, we could not rely on full and complete race reports being sent back to Widdrington Road with failed engines, and this was one of the reasons why eventually we attended most of the Grand Prix races in Europe, and in the end even went over to America and Mexico. With Lotus, Cooper and Brabham as our main customers, together with Lola, Reg Parnell and various private owners, Harry Spears had a busy time whenever he attended a meeting as engine consultant/mechanic. Perhaps that explains why Peter Windsor-Smith and I were pressed into service several times as engine rebuilders!

Back at Widdrington Road I had a small development department under Gray Ross, with no more than three fitters, who were kept busy improving the engines for the future, and I always fought hard to keep them separate from Harry Spears' production build and service shops. But during times of extreme stress it had to be 'all hands on deck' to make sure we met our dates, and there was a bit of cross-fertilisation at times.

The story of how we changed from carburettors to fuel injection, in 1963, followed by shortening the stroke, using 'flat cranks', and again shortening the stroke in 1964, has been detailed in many motor racing history books. It was Reg Parnell who first urged us into considering fuel injection; he was keen on the Tecalemit layout, but we decided instead to adopt the Lucas system. By comparison with Weber carburettors, the power output of an injected engine was no higher, but on the circuits we found that there was much better response out of corners, and the drivers much preferred the injected engines for this reason.

All the time we were intent on squeezing more and more power out of that V-8 so that we could keep ahead of BRM and Ferrari; towards the end of the formula we were also attacked by Honda, whose engines produced a great deal of power – probably more than ours could – but in chassis which were never as effective as the British machines. They only managed to win one $1\frac{1}{2}$-litre race.

During 1963, Peter Windsor-Smith and I decided that the way to real power increases in the FWMV was to shorten the stroke considerably, and allow the engine to breathe more deeply at higher rpm. We did not achieve everything at once, because I liked to move slowly and with some certainty, and consequently our tuning modifications were carried out in three or four stages. At the end of 1963 we settled on the largest cylinder bore so far, and decided to design a four-valve cylinder-head. Technically this was the correct thing to do, although very few racing engines with such a layout had been used since the war, the only recent four-valve engines at that time, I believe, being the Borgward, which was used fairly successful in Formula Two Coopers driven by Moss and Bonnier. The Offenhauser engine of Indianapolis fame had always used four valves, and Ford of Detroit had also decided to go the same way for their Indianapolis twin-cam, though this was not apparent until our own engine was well on the way.

Putting a four-valve head on the FWMV meant a pretty major redesign, for we had to use new crankshafts, pistons and liners, and a gear drive to the cylinder-head. As we raised engine speeds on the V-8s we had encountered more and more trouble with the original chain drive to the four camshafts, so a change to geared drive was logical. We had the four-valve engine running by the beginning of 1964, at the same time as we started on an even more ambitious project, which is described later, but for months we could not find the improved power on the test beds that we were sure was available. We were getting very browned-off with seeing results which were no better than those already achieved on the two-valve versions and it was not until late summer when Peter and I decided to take a hand ourselves, that the problem was finally solved.

During the summer holidays, when the people in charge of the test houses were away, Peter and I started running the engine. Already it was known that the exhaust stacks were getting much too hot, which is usually a sign of incomplete combustion, with final burning taking place downstream of the exhaust valves. We studied all the previous test-bed power curves and noticed that whenever a change was made which increased the power produced, the exhaust pipes grew ever hotter. We then discovered that it was this phenomenon which had caused our testers to try out methods to reduce this temperature which actually reduced the output in the cylinders. There also appeared to have been a mental block somewhere, because in spite of repeated proddings from Peter and myself the testers, somewhat obstinately, had been reluctant to advance the ignition beyond what was considered a normal limit. That was it! We advanced the ignition settings considerably, and with a little more attention the engine began to out-perform the two-valve version. After this breakthrough we never looked back.

Unhappily, developing the four-valve version of the FWMV had dragged on so long that we missed out in the 1964 season altogether and even missed the first few races in 1965. However, using the ultra-short-stroke configuration with slightly modified two-valve heads gave us a useful interim engine and slightly more power than in 1963. Once we put the four-valve engine into Jim Clark's Lotus 33 there was no doubt that it was very raceworthy. The interesting paradox here is that we supplied two four-valve engines – one for Jim Clark's Lotus and the other to Jack Brabham, who was now building his own Formula One cars. Although Jim's engine was slightly less powerful than that supplied to Brabham, it had slightly better torque (that should take care of the 'special engines for Jim Clark' stories), but somehow Jack never came to terms with the 'feel' of his engine, and it was only raced at Monaco early in the year.

From time to time, modifications – some major and some minor – were introduced for our Formula One engines and offered to all our customers, and nearly all engines were updated from time to time, although the four-valve conversion came along too late to be applied to customers' own engines. Most users converted to short strokes and fuel injection for 1963, but after that *some* engines were converted to flat cranks, *some* were converted to the ultra-short-stroke configuration, and all received the minor development changes Gray Ross' department produced from time to time. Every time we carried out major updating exercises, we made rebuilding facilities available, and of course we had to charge for this.

We tried to be fair and reasonable in regard to fees, and for the inevitable repairs and rebuilds which had to be completed between races during a busy season. Some customers thought these charges high, but the more knowledgeable had a good idea of the amount of time and effort we put into engines for motor racing and we had few complaints. The cost of a new FWMV, in 1962, was £3,000 as I have already said, but it probably cost more than £1,000 to have a major updating conversion carried out. That cost, naturally, depended on the scope of the work. In the winter of 1962/63, when we offered Lucas fuel injection, short-stroke cranks, revised cylinder bore details, new pistons, liners and the rest, we did it at a package price of £3,000. At the time we were already thinking of four-valve heads, and without revealing this in detail we also quoted a further price of £2,000 for 'a new type of cylinder-head, if and when produced'. All-new engines in 1963 and 1964 were priced at £5,000 each. At no time did we cover more than the cost of the parts when making these charges. The original proposal in 1964 was that the four-valve engines should be sold for £5,000 each, which certainly would not have covered the cost of design, development and build; in the event, of course, we did not sell any for the two engines built were retained by the company and loaned to Lotus and Brabham.

Mr Lee was well aware of both the advantages and the financial penalties of motor racing, which must have hit his own pocket (as chief shareholder) very hard at times. With us it was certainly a case of 'all or nothing', and if we were to be involved in Grand Prix racing we had to be sure that our engines were kept at the forefront. It was technically exciting and very satisfying when we won (or rather when drivers using our engines won) but there were adverse effects in the meantime on the extensive industrial work that my colleagues and I were trying to complete. Even so, it must have come as a great shock to everyone in motor racing – particularly the teams using our FWMV engines – when Mr Lee had to put out this announcement in October 1962:

"We have to inform you that we have decided to withdraw from Grand Prix racing at the end of this year, as it is no longer economic. Commitments we have outstanding with owners of our engines used for racing will be fulfilled. For those owners of the V-8s, which have won four first places in seven races held so far this year, we shall if required provide service after the South African Grand Prix, and thereafter supply spare parts if the owners decide to race them next year.... It is, of course, our sincere regret that we shall not be taking part next year."

To the teams who had only just become used to their V-8s, and who were already looking forward to the 1963 power-increasing modifications that we had under development, it must have been a shattering announcement, for 1962 had been their first competitive year under the $1\frac{1}{2}$-litre formula. Right away the constructors started talking, persuading, cajoling – in fact they tried every way they knew to persuade Mr Lee to change his mind. But with so much money being spent on Grand Prix racing and only a smallish proportion coming back to Coventry Climax I could not have blamed him for sticking to his decision to withdraw.

Fortunately for everyone, there *was* a fairy godmother to save the day – or rather I should say that there were several fairy godmothers. For weeks after our shock announcement, Lotus, Cooper, Lola and others talked hard and persuasively to their sponsors – notably the fuel and tyre companies – and just in time to make it a Happy Christmas for everyone Mr Lee was able to put out

Jim Clark, driving his Lotus to victory in the 1964 British Grand Prix at Brands Hatch, was by far the most successful user of Coventry Climax engines in Grand Prix racing. He scored no fewer than 19 victories between 1962 and 1965

Who else but Colin Chapman on the fork-lift truck and Jim Clark driving him? I am wondering whether we should leave him up there until he is prepared to pay more for his engines

this statement:

"Following my statement on October 17th that my company intended to stop the manufacture and development of Formula One racing engines, I was invited to discuss this decision with some of the leading members of the British motor industry. At this meeting it was emphasised what great importance they attached to Grand Prix racing as a testing ground for new motor engineering developments.

It was proposed that the motor industry would increase its financial support to a limited number of Formula One racing car constructors so that they, in turn, could contribute at least in part to our development expenses. Recognising the importance of the facts presented to me for the reconsideration of our decision, we have agreed to produce a limited number of Grand Prix engines next year. These will incorporate the lessons we have learned during the present season...."

After that rather depressing couple of months, we breathed a sigh of relief and settled back to the non-stop process of chasing reliability and power. Incidentally, that extract from Mr Lee's statement about 'Grand Prix racing as a testing ground for new motor engineering developments' was having an immediate effect on our work! Unbeknown to us, Colin Chapman and Lotus had come to an agreement to use Esso fuels and oils, and for reasons quite unconnected with the normal worth of the oil, Esso lubricants were unsuited to the FWMV. Lotus had been experiencing a spate of tappet failures in their engines, and it was some time before we discovered that their racing lubricants were different from those we normally used, and from those used by other customers. To kill the problem we eventually resorted to using some of Baron Beck's excellent Molyslip compound, which we buffed into the tappets very thoroughly to provide a satisfactory surface treatment. I used to send people out to the local Halfords' shop to buy the stuff, and it was only a chance encounter with Baron Beck which let him know about this. Naturally enough the good Baron used this knowledge in his motor show advertising in the autumn, and there was a considerable rumpus between ourselves and Colin's sponsors.

These days racing car constructors are very honest with their sponsors. Even if there is a slightly better product than that which they are contracted to use, substitution is rarely practised. Now in my Brooklands days it was very different. More than one well-known racing driver signed up with very minor oil companies for a suitable fee, but you could be sure that the oil cans he carried ostentatiously around the paddock would be full of Lord Wakefield's excellent Castrol 'R'!

As far as Coventry Climax were concerned, this public worry over the cost of Grand Prix racing was only the first straw in the wind, and it was to occur more regularly in the next few years. If there hadn't been another major change in Formula One, which was first announced in the autumn of 1963, two years before it was to become operational, the company could perhaps have carried on building and supporting Grand Prix racing engines much longer than it did, but we simply couldn't face yet another complete design project for 1966. But before then we had already decided to replace the FWMV V-8 engine by a more powerful design. Looking back, I take pride in the neat little power plant we evolved. It was only time which caught up with us before we could race it, but its development is interesting enough to warrant separate analysis.

Coventry Climax FWMV V-8 engine

As already related, the bare bones of the FWMV $1\frac{1}{2}$-litre V-8 engine evolved from a developed outboard-marine power unit which had started life at 653 cc with a single overhead camshaft, progressing through automotive applications at 745 cc, and finally to a twin-cam racing version, the FWMC, for Lotus to take to Le Mans in 1961. To clear up the conundrum of names, the original outboard was the FWM (featherweight marine), its motor industry equivalent being the FWMA (FWM automotive). The twin-cam version was the FWMC, (FWM twin-cam), and the Formula One engine's initials, FWMV, stood for featherweight marine V-8. Not a very romantic explanation, but typical of the way projects were coded at Coventry Climax.

In its original form, the V-8 engine was first raced in Jack Brabham's Cooper at the Nurburgring in August 1961, where it distinguished itself in practice first by shearing the distributor drive shaft, then by helping 'Black Jack' to a front-row grid position for the race; in fact Brabham's lap time was 6 seconds slower than Phil Hill's V-6 Ferrari and only 0.3 second quicker than Stirling Moss' FPF-powered Lotus 18/21. Brabham led the race from the start, but went off the road on the first lap due to a wrong choice of tyres.

Two engines were out on loan for the Italian GP in September, but neither Moss nor Brabham could overcome water-loss problems; Moss' car did not start, and Brabham's soon retired. However, things went better for the United States GP in October, where Brabham's engine had the modified liner-support system. Moss elected to use his FPF-engined car, but Brabham swopped the lead with him for 59 of the 100 laps before his engine succumbed to overheating once again. However, the problem was finally solved back at the factory, so the 1961 American race was really 'the end of the beginning'.

For 1962, apart from the first two engines, which were withdrawn from Moss and Brabham, rebuilt and subsequently re-loaned to Lotus, a further 16 'production line' FWMVs were laid down. Thus there were 18 engines in use in 1962 (including the two retained temporarily by the factory for development work), the last being delivered in August. For the record, three each went to Lotus, Cooper,

Yeoman Credit and British Racing Partnership and two each to Brabham and Rob Walker.

Right from the start (when Jim Clark's Lotus 24 beat Graham Hill's V-8 BRM at Snetterton in the Lombank Trophy race) the record looked good. In Grands Prix there were four wins, five seconds and four third places, with all major competition coming from BRM. It was Graham Hill's World Championship year, and it became clear that the BRM V-8 was fully competitive at the time. The FWMV's original power output of 186 bhp was enough to see off the V-6 Ferraris, but for 1963 a developed version was needed to beat BRM. Lucas fuel injection was the cornerstone to this work, with a new short-stroke configuration and, later, a 'flat' crank. Development then proceeded steadily through a series of 'Marks', the original unsuccessful 1961 engine being Mark I and the 1966 engine the Mark X. Details of the various stages of FWMV engine development are listed in an Appendix at the end of the book.

From 1963 to 1965 the FWMV powered 18 Grand Prix winners and helped to win many minor races. Remarkably, no fewer than 16 of these are credited to Jim Clark in the Lotus 25 or 33, this driver-car combination being so superior to any rival that Clark nearly always won if his car survived a race. In the same three years, FWMVs also notched up 10 seconds and 15 thirds, several times achieving 1-2-3 results.

Major changes over the years were almost entirely internal. The short-stroke 1963 engines were followed by some ultra-short-stroke units for 1964, and two of these were converted to a four-valves-per-cylinder layout for 1965. Inlet-valve size crept up from 1.3 in to 1.4 in, and the ultra-short-stroke engines had fatter exhaust pipes. Fuel injection was tried first on the very oldest engine – the one first tried by Jack Brabham in 1961 and subsequently loaned to Lotus.

Coventry Climax records show that eight of the 1962 engines were converted expensively to the full short-stroke, fuel-injection layout for 1963, and a further six new engines were built at the same time (two each for Lotus and Cooper and one each for Brabham and the Reg Parnell team). For 1964, many existing engines were further updated to the ultra-short-stroke Mark IV specification, and a further four new engines were built, all for loan (to Cooper, Brabham, and a pair for Lotus).

Finally, for 1965, a pair of Mk VI / VII four-valve engines were loaned to Jim Clark and Jack Brabham. Brabham's engine, though slightly more powerful than Clark's, never won him a race, while Clark won five Grands Prix with his. In total, Coventry Climax can identify 33 1½-litre FWMV V-8s of various specification used in racing cars between 1961 and 1965, plus several un-numbered prototypes kept for factory development use.

When the 3-litre formula arrived for 1966, all the FWMVs became redundant, and Hassan refused to consider possibilities of supercharging; there was no subsidiary formula to which the engines could be diverted. However, for 1966 British constructors were as far behind Ferrari as they had been in 1961, and there was no replacement engine. BRM enlarged their engine to 2-litres, and later to

2.1-litres, but only the persuasive Colin Chapman could persuade Coventry Climax to do the same for him. For Team Lotus only, a two-valve Mark IV engine was enlarged to 1,974 cc by combining the final cylinder bore with the original '1962' stroke. With 1½-litre cylinder-heads, ports and valves, no new machining and virtually no other development, 240 bhp at 8,800 rpm was achieved. A second engine with 1.40in valves produced 244 bhp at 8,900 rpm. Although only a stop-gap, the 2-litre FWMV won its first race in Mike Spence's hands (the non-championship South African GP on New Year's Day, 1966, which was actually the first race to be run under the new formula). Later in the season Jim Clark battled hard and often with more powerful Ferraris and Brabhams, his best placings being third in Holland and fourth in the British event. The old design still refused to lie down, for Clark then took his over-worked Lotus 33 'down under' where he won the 1967 Tasman series outright! Three 2-litre engines were eventually built, Clark using them all from time to time. Later, Donald Healey borrowed a pair for his unsuccessful Healey-Climax Le Mans car, and John Coombs borrowed one for Mike McDowell's hill-climb Brabham.

Two minor details should be explained. The flat-crank engine was not developed after long-considered research, but to ease engine installation in a car which was never built! Ferguson Research wanted to follow up their interesting FPF-powered P99 four-wheel-drive car, and designed a car with a front-mounted FWMV, which meant that the familiar 'spaghetti' system could not be used because the tailpipes would have had to pass through the driver's body! Hassan, therefore, was forced to investigate layouts which could retain outlet pipes from each bank without a cross-over, and the flat-crank was the only way that suitable exhaust pulses could be obtained. Power output levels were maintained, and there was no noticeable increase in vibration, which theoretically should have occurred. The revised exhaust system was so much more simple to install in a car that many engines were rebuilt to take advantage of this in 1963 and 1964.

The ultra-short-stroke engines (1964 and 1965) were not developed primarily because of their high-revving capability, but rather to provide a larger-diameter cylinder bore to accommodate the movements of closely-packed valves in the four-valve head then under development. Hassan's narrative has already made it clear that the four-valve layout was delayed by combustion problems, and the two-valve engines tried as a stop-gap worked very well.

The very successful FWMV was not, of course, the last racing engine designed at Widdrington Road. The FWMW flat-16 - that intriguing little 1½-litre design which ran out of development time - came last. Like the first Coventry Climax twin-cam, the FPE, it was never installed in a car although it achieved a competitive power output. If the formula had continued beyond 1965, and if Coventry Climax had decided to continue racing, the FWMW would certainly have made its mark. How and why it came about is the subject of separate analysis in the next chapter.

8
The stillborn Sixteen

The 1963 fuel-injected short-stroke FWMV engines, many of which, in fact, were extensively rebuilt 1962 engines, were a great success. Even so, towards the end of the Grand Prix season, which was dominated by Jim Clark's Lotus-Coventry Climax, we sensed that other teams' engine manufacturers were beginning to catch up. This might seem difficult to believe after a season in which Jim Clark in taking the World Championship had won seven Grands Prix whereas BRM had scored twice and Ferrari only once, but there were many occasions when drivers other than the brilliant Scotsman were having to work very hard to stay at the front of the field.

The FWMV had developed about 186 bhp in 1962, while most of the short-stroke examples developed between 195 and 200 bhp in 1963; we had the ultra-short-stroke four-valve design under way, but we were still not sure it would suffice until the end of 1965. We were always very respectful of Ferrari and BRM engines, but it was when we heard that Ferrari had started testing a car with a new V-8 engine that I became quite perturbed. British cars and engines had had it all their own way in 1962 and 1963, but in the meantime Ferrari had designed a new car rather like the Lotus, hired a new driver and tester in John Surtees, and now had what promised to be a very powerful new engine. More than that, I also heard through the grapevine that Ferrari was working on a new flat-12 $1\frac{1}{2}$-litre which would make both of his other engines obsolete!

All this became clear at about the time the new 3-litre formula was announced. I was not sure whether Mr Lee would decide to support the revised formula (he certainly didn't make up his mind right away), but I was reasonably certain that Coventry Climax would continue in racing to the end of the $1\frac{1}{2}$-litre formula in 1965. This meant that there were two more seasons during which I wanted to stay ahead of the opposition. Peter Windsor-Smith and I studied the problem, surveyed the remaining potential we judged to be hidden inside the V-8, and considered what Ferrari, BRM and Honda might do – then decided to carry on in two ways.

First we would complete major changes to the V-8, by increasing the bore and shortening the stroke – this would give us the potential of higher engine speeds and enough space in the cylinder-head to use the four-valve arrangement. We also decided to design a completely new engine! I mention the four-valve V-8 once again because the trials and disappointments we suffered with it affected our progress with the new design. Quite a few of the features from the final V-8 were also to be applied to the new engine, but because of the maddening delays we suffered on both counts.

The new engine used no parts from any of our existing engines, though we tried to build in the mass of our experience gained since the early 1950s. I was quite sure that we needed significantly more power than the V-8 would give – an extra 20 bhp at least – and the only logical and straightforward way to achieve this would be with the aid of higher engine rotating speeds, greater piston area and good breathing. But having said that, I must point out that it really isn't simple to design such engines, which have to be the most efficient in the world. All the theoretical methods of achieving high specific outputs are known, but actually seeing these vindicated in a reliable racing engine is not at all easy!

Peter and I based our project studies on engines of various configurations, but having one common factor – we thought a stroke/bore ratio of around 0.75 to 1 seemed to be the ideal. This was the ratio which we were using in the 1963 V-8s, though the later four-valve version was to be much more over-square than that. As we were thinking in terms of multi-cylinder engines with peak revs around 12,000 rpm we also had to consider the overall gearing required by Grand Prix cars. Our first V-8s had peaked at about 8,500 rpm, although later Marks improved on this, and to allow the constructors to use existing gearboxes and axle ratios we decided to use a geared final power shaft so that the transmission would still receive its peak power at around 9,000 to 9,500 rpm.

It would have been very nice to start quite literally with a clean sheet of paper, but this did not happen; in my experience it very rarely does! We had to come to terms with the fact that any new engine we designed should be capable of fairly straight-forward fitment into Grand Prix cars which had already been built – we could not expect Lotus, Cooper and Brabham to design completely new racing chassis to fit around our latest efforts. Right away this virtually defined a certain volume or 'package' inside which we had to accommodate a new unit, a package complicated by the fact that while Cooper and Brabham were still loyal to multi-tube chassis frames, Colin Chapman's Lotus had a 'bath-tub' shape of monocoque with bulky sponsons running under the sloping heads and blocks of our FWMV V-8s.

I decided that 220 bhp would be an adequate minimum power target at first. The engine would have to be a simple design to satisfy my own standards, it should not be over-expensive, nor should it contain too many untried features. Given the same engine capacity limits and fuel specification, we could obtain such power by an increase in operating rpm, or by increasing the piston area, or of course by both. As the late L.H. Pomeroy once said in a famous paper on engine design: "Revolutions are of an abstract nature, they cost nothing, weigh nothing, have no shape or substance. If one can get more revolutions than another, it is difficult to find a reason why he should not benefit thereby." I agreed with this, and quoted Pomeroy in a paper on racing engine design to the SAE in the 1960s. Further, I also said: "Whoever can make his engine revolve faster than his competitor, and also arrange for it to remain in one piece, will succeed in obtaining the highest power, always accepting a condition of satisfactory aspiration."

Piston area, or rather its increase, is also a known factor in power improvements, so a consideration of this and the rpm factor - and of the installation benefits - showed that of our only reasonable options of 12-cylinder or 16-cylinder designs, the 16-cylinder layout had the greater potential. This was also confirmed because we were wedded to the theory that the most efficient exhaust extraction conditions are obtained by coupling four cylinders firing at 180-degree intervals; these are far in advance of coupling three cylinders together, as is necessary with a 'twelve'.

The 12-cylinder layout also seemed to offer little in the way of frontal-area reduction if we opted for a conventional 60-degree vee formation. However, there was one 12-cylinder layout to which I was attracted, and I still feel it to be a good one. This would arrange the cylinders in three banks of four each, laid out in the shape of a 'broad arrow' like the very successful Napier Lion aero-engines which found motor racing fame in Sir Malcolm Campbell's Bluebirds, in Golden Arrow and in the Napier-Railtons. We didn't pursue this layout seriously in 1964 because there promised to be complications in big-end and crankshaft design, but it is still a basic configuration that I would like to see adopted for a future racing engine.

So it was to be a 'sixteen', but what form of 'sixteen'? In our view it could only be built in one of the many possible variations of two banks of eight cylinders. We thought briefly about a four-bank engine layout which Rolls-Royce and Napier had pioneered with some success in large military aero-engines, but I was always suspicious of the layout. Theoretically it might be very compact, but there would have been many worrying unknown problems involved in the coupling together of two crankshafts. I must say I thought it was very brave of BRM to build and race their H-16 3-litre Grand Prix engines in 1966 and 1967, but although they could guarantee a competitive power output they suffered from all the problems we had foreseen a little earlier. A large company with nearly unlimited resources could, perhaps, have tracked down the built-in problems I had anticipated would occur, but Coventry Climax would not have been able to cope.

For balance purposes, a 16-cylinder in two banks of eight could be a wide-angle V-16 like the original supercharged BRM, could have a slightly drooping layout (with the vee angle pointing downwards), or it could be a completely horizontally-opposed engine.

We sketched out all these units and investigated them seriously. The wide-angle V-16 was compact, but with a power take-off shaft intended to be below its crank line we would have been faced with a higher centre of gravity than was desirable. The 'drooping' layout offered a very acceptable centre-of-gravity position, with the power take-off shaft nicely tucked away, but getting an exhaust system into the cars would have been very difficult, and we would have had problems in clearing the lower chassis frame tubes in the Brabhams. Eventually we settled on the flat-16 layout, even though this meant that the Lotus monocoque would have to be altered substantially.

The constructors appreciated our care in this investigation, especially when they were shown the theoretical advantages of a 16-cylinder engine. We had already achieved more than 200 bhp at 10,000 rpm with the V-8s, and if we could achieve the same breathing and combustion standards in the 'sixteen' we might see something like 230/240 bhp at 12,000 rpm. That would be a dramatic leap forward – almost as startling as the improvements we made in 1962 when we replaced our old four-cylinder FPFs with the new V-8s. To find 240 bhp would be the equivalent of achieving 160 bhp per litre, a very ambitious target which, incidentally, was still not being attained consistently in Formula One engines ten years later.

The layout and detail design of the FWMW, as the flat-16 was designated, is well known, but I doubt if many people realise that we intended the little engine to have four valves per cylinder. When we built the first four-valve V-8s early in 1964 I was intending that lessons learned in this engine should be passed on to the other, and we certainly designed our first FWMW accordingly. Looking back on it, this was not as complex a mechanism as it might seem. Each of the four cylinder-heads would have accommodated sixteen valves, and since the whole thing was meant to be built up like a very sophisticated Meccano set it didn't look too frightening. However, we reached the point of no return with the design where we had to commit it from paper to actual components before we had made sense out of the original four-valve V-8's combustion, and so reluctantly I decided to revert to a classic two-valve layout.

By now, in fact, time was beginning to catch up with us in many ways. The prototype flat-16 was not completed until the end of 1964, and with only one season of 1½-litre racing to look forward to it was obvious that the 'sixteen' would have to work immediately if it was to be a success. Certainly we didn't intend to make many – at the time the engine was shown to the press in February 1965 we had decided to build only four, one each for our three principal customers, and a spare to be kept at Widdrington Road for development and for use in emergencies. Each of the constructors set about modifying cars to accept the 16-cylinder engine, which was just one inch longer than the race-winning V-8, slightly narrower and several inches lower, though our cylinder-heads would poke into the space normally occupied by the Lotus rear sponsons.

The FWMW was an exciting-looking engine, and we couldn't wait to prove

The 1½-litre FWMW flat-16 Grand Prix engine, for which we had high hopes when we designed it at the end of 1963, but for which time eventually ran out

Harry Spears explains details of the FWMW flat-16 engine to HRH Prince Philip, the Duke of Edinburgh, during his visit to the Widdrington Road factory

that it worked. One well-known motoring writer described it as: "A sensational design by any standards . . . it is probably the most advanced unblown engine yet built in this country" – which was, after all, what we had set out to achieve.

Even though we were running desperately late in the life of the formula we were very optimistic. However, the very first time we fired it up it blew up! We had completed all the usual pre-running checks, then started it up in the normal manner, whereupon it idled satisfactorily at about 2,000 rpm. Unfortunately we simply couldn't persuade it to run any faster, even by using all our experience in coaxing temperamental and highly-tuned units to gain speed. While we checked mixture strengths, fuel flows and so on, which took about five minutes, smoke began to pour out of the breathers, and suddenly the quill-shaft, the final-drive take-off from the centre of the crankshaft, sheared! That was instant disaster as far as we were concerned, but we were soon able to pinpoint the problem. We were in a severe torsional vibration problem at that speed – low-speed pulses had caused the shaft to vibrate, and with the high inertia of the dynamometer holding down the other end it heated up and broke.

We set about obtaining a stronger quill-shaft, but this took several weeks which we couldn't afford to lose, then to make sure it never happened again I issued instructions that the engines should not be allowed to run at speeds below 4,000 to 5,000 rpm. We soon achieved reasonable reliability, but then we ran into the most worrying of any racing designer's problems – the engine simply would not deliver the power it should have done! This time we were quite sure of our forecasts on the breathing, and we knew that things like ignition timing and fuel injection settings were right, so it soon became clear that the discrepancy was due to internal losses. I must make it clear that the flat-16 was not dramatically down on power, in the way that the first V-8 had been, but it was not showing the immediate improvement over the V-8 I had expected. If we had put one straight into a car I believe it would have been competitive with anything then installed in a Ferrari or a BRM, but so was our four-valve V-8, and I wanted to achieve much more than this.

The work we did to trace the internal losses taught us much about really high-revving and complex engines. There were oil-drainage losses, pumping losses, windage losses, all of which could certainly have been solved if only we had had time. As it was, we achieved 209 bhp on occasion, which was just about on a par with Jim Clark's championship-winning four-valve V-8, and about six bhp better than we ever saw on the older versions, but it still wasn't enough for us to consider letting any engines be fitted to cars.

It was a disappointment in what was destined to be the last season for our racing department in Coventry, but I was happier to go out on a winning note with our old V-8s than be remembered for a final unsuccessful effort. By the middle of 1965 we had decided to fold up the project, put the single engine and all the spare parts in a corner, and forget all about them. They joined a very small pile of work on other engines that had been tried, or proposed, but never brought to fruition, notably the 1-litre Formula Two design we made around the old FWA block for possible use in 1964 and after.

There was an amusing sequel. In 1966, Coventry Climax received a visit from HRH Prince Philip, Duke of Edinburgh, and Mr Lee decided that we should put on a really outstanding technical display for him; we were to bring the flat-16 FWMW out of retirement, and run it up on the Widdrington Road test-bed for his benefit. This we did, and when Prince Philip reached the test house the engine was well warmed up and howling away splendidly. Unfortunately, this was followed by what we used to call 'expensive noises' and the engine came to a sudden and grinding halt! I am sorry to say that such was the pressure of work on non-racing matters by then that we never stripped the engine to find out what had happened. The engine, I believe, is still in a 'blown up' condition, and I don't suppose we will ever know what caused it!

Unfortunately, when we had announced the flat-16 engine in February 1965, we had also been forced to put out a statement to the effect that the company proposed to pull out of motor racing at the end of that year. Unlike 1962, when Mr Lee's original decision to withdraw was reversed in a matter of weeks, this time there could be no retraction.

There had been plenty of warning about a change in formula, as the new 3-litre Grand Prix limit had been announced at the end of 1963. Our attitude to all this was precisely the same as it had been in 1958 when we were faced with a similar change. However, it was fascinating that while the 1958 changes had all been intended to cut speeds and reduce hazards (which they failed to do for very long), the new proposals doubled the engine size and would certainly make the new cars the fastest ever. The promoters had called for greater spectacle, and they were certainly to get it.

In our view these changes were so illogical, especially in view of the constant complaints about the exceedingly high cost of power units, that we felt unable to support them. The design of a completely new power unit of twice the size, with all the development costs, was more than we could swallow, especially as we had been led to believe from the constructors that they would not pay up! Had the formula not been forced to change, I have no doubt at all that we could have carried on much longer, with a developed flat-16.

We certainly didn't withdraw because we had lost interest in the motor racing scene. Mr Lee accepted the recommendations made to him by other directors and myself, which touched on the change of formula and the mass of commercial work which we felt was in danger of neglect. I think that in his heart he wished we did not have to withdraw, and on several occasions since then he has told me how sorry he was. However, I want to make it clear that there was no influence from Sir William Lyons, who by then was Chairman, following the company's acquisition by Jaguar, and as long as Coventry Climax engines had remained successful in Grand Prix racing I am sure he would have been content to let us continue.

Once we had made our decision to pull out, and informed the constructors and the press, we planned a clean break. Peter Windsor-Smith and I never even got round to sketching out 3-litre layouts, although we had a very clear idea of the sort of engine we would have designed. When the new 3-litre unsupercharged limit was announced, there was an option of using supercharged 1½-litre units. On the evidence of reliable outputs achieved by people like Alfa Romeo in the 1940s and 1950s, a supercharged 1½-litre, *designed as an entity*, could have been very powerful and competitive. I need hardly say that as a stop-gap I was approached several times regarding the possibility of supercharging the FWMV V-8, or even the unraced flat-16! The approaches came from people who were hoping that they would not have to buy all-new engines, or to build all-new cars. However, even though supercharging doesn't necessarily mean over-stressing what was originally an unblown engine, I was quite convinced we could not do the job properly, and advised against it.

The basic reason was that we couldn't think of where to find a suitably small turbocharger - because it would have had to be a 'turbo' rather than an engine-driven blower. At the time we were looking around for small turbochargers for our industrial diesel engines, but were having little success, even in the United States; nowadays the situation is much improved because Holset market an American device in Britain.

I have said that our decision to withdraw from racing was final and that there was a clean break, but there was a single instance where we broke this resolve. We built a couple of 2-litre FWMV engines to help Lotus and Jim Clark in 1966, as Lotus were unready for the new formula in spite of having more than two years' warning of its coming. They had assumed that we would continue racing under the new 3-litre limits, and were caught without alternatives at the beginning of 1966. We felt able to help Jim Clark in the interim because, along with Coventry Climax, he and Lotus had won a remarkable number of events under the old 1½-litre rules. There was, however, very little new development involved in this 'interim' engine; purely by chance we discovered that if we combined the biggest cylinder bore ever used and the longest stroke ever used in the V-8s, the resultant swept volume was almost two litres. So we had a situation where 1965 bore + 1961 stroke = 1966 engine! Once they were built we did no further development. Even so, Clark achieved a number of good places during 1966, and went on to win several events in Australia and New Zealand to take the Tasman Championship at the beginning of 1967.

The last Grand Prix which we supported fully was in Mexico in October 1965. It would be nice to record that we went out on the crest with a win, but this turned out to be the only 1½-litre Formula One race to be won by a Honda; Richie Ginther's transverse-mid-engined car beat Dan Gurney in a Coventry Climax-powered Brabham by less than three seconds.

The 3-litre formula came into force on January 1st, 1966, and immediately rendered all the old FWMVs obsolete. Almost at once, apart from Jim Clark's 2-litre Lotuses, the old cars disappeared, and with them the engines. But unlike snow in summer, they didn't melt away to be lost for ever. Most of the cars which survived were preserved in one form or another, and several found their way to the splendid array of single-seater racing cars owned by Tom Wheatcroft, and are now displayed at the Donington Collection in Derbyshire. In addition to Coventry Climax engines installed in various cars there, one of our later two-valve FWMVs is displayed on a separate show stand.

Mr Lee used to enjoy his motor racing, particularly when meeting the cosmopolitan GP circus. We attended as many of the races as possible, though at first we had acute difficulty in arranging passes from the organisers. They treated us like outsiders, because we were not actually racing, but by dint of scrounging passes – one from Lotus, one from Cooper, and so on – we always managed to get there in the end. Mr Lee's speciality was in arranging impromptu dinner parties to celebrate a victory, and those at the Hotel de Ville in Monza were usually the liveliest! Colin Chapman usually started the bun-fight and had a good aim. On one occasion we were entertaining the son of one member of the Royal Family who, sitting between Harry Spears and myself, was really anxious to get involved. Once we had supplied him with suitable missiles – bread rolls soaked in water – there was no stopping him, and it was only when he decided to bring soaked serviettes into the line that we had to call a halt. I gather it took some time for his family to persuade him that a bun-fight was not a normal function of a celebration dinner!

There is something about the atmosphere of motor racing which gets you; it certainly held me, and I always try to make an annual pilgrimage to one Continental event, usually Monaco, to keep abreast of things. There is much commercialism about these days, of course, but still room for kindness and thought. One of the nicest gestures I have ever witnessed in motor sport came about some time after we had withdrawn from racing. Colin Chapman acknowledged the debt Lotus owed to Coventry Climax for the fine engines which had helped him to so many Grand Prix successes, and he presented the last of his Lotus 33s – complete with its Coventry Climax engine – to Mr Lee as a souvenir! Of course, Mr Lee was delighted and rather touched by this, and it was completely characteristic of the man that he, in his turn, presented the car on long-term loan to Coventry's Herbert Art Gallery collection of historic vehicles. Later still, of course, the car found its way to Tom Wheatcroft's collection at Donington Park, where it stands alongside the engine I have already mentioned.

As to the flat-16s, at the time of writing none of these is on show in public. When Mr Lee retired and handed over complete control of the business to Sir William Lyons, he took with him examples of all the successful racing engines, including that flat-16 which had been 'blown up' for Prince Philip's benefit, and the spares to make it go again. But as far as I know this rebuild has yet to be carried out.

As I write, almost a decade has passed since Coventry Climax built its last Grand Prix engine, but I am happy to see that they are still a famous name in motor sport. From time to time the motor racing press invents stories about

The Coventry Climax luncheon table for Prince Philip's visit. Miss Morris is on Prince Philip's left and Eric Buxton is beside her, while Frank Cotton is sitting alongside Leonard Lee, on Prince Philip's right

Jaguar's engine design team in the mid-1960s. From the left: Harry Mundy (Chief Development Engineer, later to take my old job), myself (Executive Director—Power Units), Bill Heynes (Technical Director and Vice-Chairman) and Claude Baily (Chief Designer—Engines)

British Leyland building new, secret and exciting engines for a new generation of racing cars. Sometimes I think these stories have been inspired by people who are hoping to spark off some sort of programme again, but there has never been any serious intention to get back into the sport. It is true that while the late Derrick White was employed at Jaguar, before he went on to design the 1966/67 Cooper-Maseratis and John Surtees' Hondola (as the Lola-inspired Honda was dubbed), there were one or two paper schemes, but they never came to anything. As far as Coventry Climax were concerned, once we had run down work on the existing V-8s and flat-16s we never started up again. The team which achieved so much between 1955 and 1965 was allowed to disperse, each man to take up the next stage of his career. I am happy that many of them have since achieved even more notable success.

For my part, the end of our motor racing programme was really the end of an era, for I had been closely involved with motor racing and racing cars since 1920. However, the run-down certainly didn't mean that I suddenly had a lot of spare time, for now we had to concentrate on all those 'bread-and-butter' projects – for fire-pump, military, outboard and other uses – which had continued strongly and had subsidised our motor racing activities over the previous ten years. In the meantime, Coventry Climax had joined forces with another famous Coventry concern, and this was to re-unite me with several good friends and old colleagues. It had been early in 1963, soon after we had resumed our V-8 racing programme, that those two great individualists and enthusiasts for their own companies, Mr Leonard Lee and Sir William Lyons, had agreed to merge Coventry Climax with Jaguar.

Coventry Climax FWMW flat-16 engine

Fifty years hence, no doubt, a motor racing historian will call the flat-16 FWMW engine a flop, because it never raced and didn't produce superior power to the FWMV V-8 it was supposed to replace. But this judgment would be wrong, for Hassan has no doubt that it could have been made competitive if enough time and interest had been available. But the fact that Jim Clark, in a Lotus 33, with a four-valve V-8 engine behind him, was well on the way to the 1965 World Championship by mid-season, coupled with the decision of Coventry Climax to pull out of racing at the end of the year, meant that the project had simply run out of time.

By all previous (and subsequent) motor racing standards, the flat-16 FWMW was unique. No flat-16 engine has ever raced; indeed, the only 16-cylinder engine which ever graced a consistently successful racing car was in the massively brutal Auto-Unions built between 1934 and 1937. Two other 16-cylinder engines have been raced in Britain – the ill-starred V-16 supercharged 1½-litre BRM of 1950 to 1955, and the H-16 3-litre BRM, which was unsuccessful in cars from its own stable, but which powered Jim Clark's Lotus 43 to a single victory in 1966.

The engine could have been in V, H, or flat configuration, and Hassan quickly eliminated the first two options. A V-16 like the post-war BRM or the Auto-Union would have been too bulky, and he quickly discarded the H layout because he could not see the wisdom of having coupled cranks and all the vibration implications this would bring. (His views were to be vindicated by the trauma suffered by BRM with their own H-16 layout in 1966 and 1967.) The remaining choice was the flat-16, or more accurately a very-wide-angle 16 which most conveniently had an included angle of 180 degrees. In fact, a 135-degree layout was investigated (centre of gravity too high), as well as a 225-degree layout with drooping banks (difficult to install in a car).

A Meccano-type of assembly was chosen, with split block/crankcase assembly, four cylinder-heads, four exhaust systems, and a central power take-off geared to produce the same order of output shaft speeds as those experienced with the FWMV V-8. By almost any standards the bore and stroke – 54.1 mm and 40.64 mm, respectively – were tiny, and there is no record of any other Grand Prix engine having been built with such a short stroke (the V-16 BRM dimensions were 49.5 × 48.3 mm). It may have been straining incredulity too far to expect to see four-valve heads on this engine (16 cylinders, 64 valves, 128 valve springs!), but Hassan would have been very tempted to squeeze all this complication into his new engine if it had already proved beneficial on the V-8. Apart from the neat and precision-watch-like layout of the new engine its most remarkable feature was the two-piece crankshaft whose two halves were pushed together by hydraulic pressure.

Such an engine might have been bulky, but in the event it was only one inch longer than the V-8, several inches lower and marginally narrower. Virtually the only loss to the car constructors was that the flat-16 as first built was between 10 and 15 pounds heavier than the V-8, a rise brought about by the fitment of dual 8-cylinder ignition as 16-cylinder equipment was unobtainable at that time. As to its power potential, the flat-16 should theoretically have been capable of up to 240 bhp at 12,000 rpm compared with around 213 bhp at 10,500 rpm for the very best and most carefully assembled four-valve V-8, although Hassan would have been happy to see 220 bhp after initial development, which would still have given a useful margin over the latest from BRM and Ferrari, though the astronomical figures claimed for the 1½-litre Honda were thought to be out of reach.

However, because of preoccupation with V-8 developments, in particular the vicissitudes of development of the four-valve versions, assembly and testing of the first flat-16 engine took time. Design had begun as early as 1963, before Jim Clark had won his first World Championship with a V-8, but it was not until the end of 1964 that the first engine was run on a test bed at Widdrington Road.

Hassan's reaction to the delayed development of the engine are summed up by extracts from his paper read to the American SAE in October 1966: "By this time

Jim Clark had more or less got the (1965) Championship in his pocket so, regretfully, it was decided not to pursue the 16-cylinder engine. In addition, it was not sufficiently better than the V-8 to warrant the cost of running another car, even though the chassis had been nearly completed by Lotus for the purpose . . . a pity indeed we had not another year or so for development!"

9
Finale with twelve cylinders

There can be few directors of sizeable companies who hear of their takeover by reading about it in the newspapers, but it happened to me! One day early in March 1963 the local evening paper, the *Coventry Evening Telegraph*, carried the story that Mr Lee had decided to sell his company, Coventry Climax Engines Ltd, to Sir William Lyons' Jaguar group; that was the very first thing I knew about it. The other board directors had little inkling of developments, either, in fact it came as a complete surprise to all of us. In a less stable and contented atmosphere we might have been very disturbed by this move, but we all understood Mr Lee's motives. Later, he told me that he had been planning to sell Coventry Climax to a larger concern for some time, but had not carried out this intention until he could safeguard the future of his employees and staff by ensuring the continuity of the company.

At the time, there were those who suggested that Mr Lee had sold out because his company was in financial difficulties, but this was certainly not the case. Such rumours may have been fuelled by reference to the intended withdrawal from motor racing, which had been announced at the end of the previous year and later reversed. This, of course, was forced upon us because we didn't think Coventry Climax should go on losing money in its motor racing operations, and it was necessary to pressurise our customers into paying more for our products! Meanwhile, our commercial operations were strong and profitable; we were still supplying engines in large quantities (large, that is, by Coventry Climax standards) and the fork-lift truck side of the business was exceedingly healthy.

I know that the company could have been swept up by one or other of the merchant banks or insurance companies, but Mr Lee was determined to ensure the future in the way he would have planned it; although I believe he talked to more than one other company in the motor industry, it seemed absolutely logical and fitting that he should eventually get together with Sir William Lyons, who was very much a man after his own heart. Jaguar, after all, was also essentially a one-man band, and had been built up from scratch by Sir William, who had started in the 1920s and was still a very powerful and active boss in all respects. Sir William became Chairman of our company, though Mr Lee remained as his deputy and Managing Director.

Apart from the sheer sentiment of a motor racing engine manufacturer (it wasn't our main business, but that's what we were now famous for!) linking with the producer of some of the world's finest cars, it also seemed to confirm that we would be involved with fine engines for a while yet. Between 1951 and 1957 Jaguar had won Le Mans on five occasions, along with many other race and rally successes. Sir William had pulled his works team out of motor sport at the end of 1956, when he said that it was interfering too much with his commercial development and design work – a familiar enough situation! However, we also knew that he was very enthusiastic about Britain's Grand Prix successes, and we felt sure he would encourage us to keep going with our racing programme.

I must repeat that I had absolutely no fore-knowledge of the Jaguar-Climax merger, because at the time there were several other board members who thought I had had a hand in the discussions; it was remembered that Sir William had approached me to return to Jaguar several years previously, although I had stayed at Coventry Climax at that time. I was always happy about the merger, because I was sure our own engineering activities would be complementary to Jaguar work, but other members of the staff were not so sure. Miss Morris, our Purchase Director, was so confused about the prospects that she said: "I don't know whether to kiss you or to hit you!". I'm sorry to say that she could not bring herself to do the former – at least not on that occasion – but we remained good friends ever afterwards.

But what about the future and how it would affect me? The Jaguar company I had left in 1950 was much changed in 1963, although I had kept in close touch with old friends there, and in particular with Bill Heynes.

The XK engine had become a world-wide success, and of course it had been used in the 'C' Type and 'D' Type racing sports cars with great distinction. I must say it was nice to know that the engine we had developed jointly from those fire-watching sessions *had* proved to have so much potential built in, as we had forecast. At the time, of course, we had suggested to Mr Lyons that he didn't really need such sophisticated engineering detail in the engine, but quite rightly he had insisted on it.

Commercially, Jaguar had expanded mightily. They had moved out of their original premises in Foleshill to a much larger ex-Government shadow factory in Browns Lane, Allesley, once run by Daimler, and by 1963 this was already bursting at the seams. Not even the dreadful fire at Browns Lane in 1957 could stop progress for long. In the few years before our own takeover, Jaguar had also acquired the Daimler company from BSA (Daimler was just a few hundred yards away from my own office in Widdrington Road), and the bankrupt Guy commercial vehicle concern in Wolverhampton. Sir William had also involved himself in a joint venture with the Cummins Diesel engine company, to make power plants for his new range of Daimler buses and coaches.

Commercially and financially, therefore, Mr Lee seemed well advised to sell his company to Sir William. Apart from ensuring future employment at our factories, there was a lot of good sense in linking our own engine-building activities and fork-lift truck manufacture with those of Daimler and Jaguar. At the time of the merger there were no obvious overlaps of product, nor immediately any obvious use for our engines in Jaguars or Daimlers. We hoped that we could continue with much autonomy, and that Mr Lee's policies would remain unchanged.

We need not have worried. Once the first burst of interest in the takeover had died down, business seemed to carry on precisely as before. Mr Lee was still in residence in his exquisitely-furnished Victorian-style office, and no staff changes took place. In regard to engineering, although I immediately renewed a business liaison with Bill Heynes after a lapse of 13 years, there was no link-up of design or development work for the time being. The only immediate advantage was that purchasing activities were rationalised. It seemed that this was almost a 'marriage of convenience', as there were to be so few links between Jaguar and ourselves. It was almost a takeover without consequences.

On a personal note, it was to be quite some time before I became closely involved in Jaguar group engineering activities, and there is really no particular date which I could pinpoint when I became a Jaguar rather than a Coventry Climax director. Instead, there was a slow, steady and informal build-up of consultation between us, which brought me into contact not only with passenger-car engine design work at Browns Lane, but also with diesel engine development at Daimler and its suppliers. It was not until Coventry Climax had withdrawn from racing engine work that I moved my office to Jaguar's engineering division at Browns Lane. Later, of course, I became closely involved in the development of the new 12-cylinder passenger-car engine, and various diesel-engine projects for Daimler and - later - BMC.

As I have already pointed out, although we might still be famous for our racing engines when historians come to talk about the 1950s and 1960s, much of our bread-and-butter activity concerned diesel engines, and at Widdrington Road we had amassed a huge quantity of expertise that was immediately valuable to Jaguar and Daimler. In the early 1960s Daimler were in need of a new range of engines, and had set up a joint company with the American-owned Cummins Diesel concern to make units for a new range of Daimler and Guy buses and coaches. I always found it strange that the factory intended to build Cummins-Daimler engines was separate and alongside the Cummins factory at Darlington built to sell engines to other customers, but no doubt there were good commercial and financial reasons for this. I had to make one or two trips to the United States to the parent Cummins factory to discuss progress, and changes required on the new 9.6 litre V-6, and was much involved in the project technically. Later, following the Jaguar-BMC merger, I was also consulted regularly by BMC who required a new large diesel for their truck range.

Commercial tie-ups in the motor industry in the mid-1960s were complicated, to say the least. BMC had a loose link with Rolls-Royce, while Jaguar-Daimler were linked to Cummins, and to add to the choice facing BMC there was also a newly-designed V-8 diesel from Perkins of Peterborough, at that time still an independent concern. Whatever happened, BMC were planning to make the chosen engine themselves, so whether it was to be a Perkins design, a Cummins V-6, or one of the Shrewsbury-built Rolls-Royces, there would be tooling changes to be discussed. Mr Heynes should have been more closely linked in policy discussions at this level than I, but he was much more interested in pushing ahead with new Jaguar passenger car engines, and deputed me to act as consultant to BMC. It all took time, but we were just approaching the decision point – which looked like being in favour of the refined Rolls-Royce unit – when the Leyland merger took place, and put the whole thing back into the melting pot!

My first contacts with Jaguar and Bill Heynes from 1963 onwards were essentially informal. During that year I was introduced to the advanced V-12 engine with which Heynes and Claude Baily were dabbling, though there was little urgency behind the project at that time. It is worth recapping what had happened technically since I had left Jaguar in 1950.

The XK engine had continued in production, with different versions being developed. There were 2.4s, 3.4s, 3.8s, 3-litres specially for racing, and by 1963 a 4.2-litre version on the stocks for fitment to 1965 models of the 'E' Type sports car and the Mark 10 saloon. The 'E' Type, announced in 1961, was conceived in 1956 as a new racing sports car to replace the 'D' Type, and would have used a specially developed 3-litre version of the XK engine if Sir William had decided to carry on in racing.

At the height of Jaguar's racing involvement, Sir William and Bill Heynes realised that they might need a lot more power to keep up with Ferrari in the future, and their thoughts turned to a new engine design. Everyone may know that Jaguar's famous V-12 engine has been in production since the beginning of 1971, but not many know that its origins date back at least to 1954! No engines were built at that time, but I have seen files and specifications at Browns Lane which confirm this date. Graham Robson, who joined Jaguar as a graduate in 1957, tells me that XJ drawings certainly existed then, though no progress had been made towards building even a prototype for test-bed work. The six-cylinder XK engine had succeeded in winning its fifth Le Mans victory that year, and following Sir William's decision to withdraw from racing at the end of 1956 the V-12 project had been temporarily shelved.

By the 1960s, when I began my informal visits back to Jaguar, there had been a revival of interest in racing. With big companies like Ford of America becoming involved, Bill Heynes was inclined to try his luck again. Claude Baily began work on updating the V-12 schemes, with the intention eventually of having one installed in a mid-engined sports-racing car to take to Le Mans.

The design, which still only existed on paper, was virtually two XK 2½-litre engines, set at 60 degrees, with a common crankcase. There were twin-overhead-camshaft cylinder-heads closely related to those Claude had drawn in 1943, and with a swept volume of just under 5-litres it was hoped that the engine would make a new car competitive against the big Fords and Ferraris which were known to be ready for 1964.

Meanwhile, the attractions of converting the XJ engine into a quantity production unit suitable for a road car had not been overlooked, and even before the first racing engine was complete the decision had been made to productionise a version of it. In fact this was the first all-new engine project Jaguar had tackled since the 1940s (apart from a still-born 9-litre V-8 for military use) and it needed the full-time attention of expert development engineers. Mr Heynes thought that a new man should be brought in to look after the project, and when I was asked my opinion about the sort of chap we should seek, I knew that my old friend Harry Mundy would be ideally suited for the job, and I had no hesitation in saying so.

As I have already recounted, Harry had left me at Coventry Climax in 1955 to join *The Autocar*, but by 1963 I thought he was just about ready to consider a move back into the industry. In the meantime, while his reputation as a technical editor had blossomed he had found time to design the twin-cam cylinder-head conversion to the Ford Cortina engine, which Colin Chapman introduced in 1963 with the Cortina Lotus and used so successfully in various Lotus products afterwards.

Although I suggested that he should be approached, and Bill Heynes agreed with me, it took a long time to persuade Harry to make the move. Although he always said that journalism was only a temporary phase in his career, he enjoyed his world-wide travels, and was somewhat reluctant to come back to a Coventry-based job. He has told me since of many occasions on which he talked to Bill Heynes, to Sir William, and even to Bob Berry, Jaguar's Publicity Manager and former racing driver, before making up his mind. He also had talks with me, and on my advice he eventually took on this exciting new task. I had thought originally that Harry would eventually become project engineer for the entire V-12 scheme, but this was not possible immediately; one has to await the proper opportunity to make a major change in staffing.

Harry eventually joined Jaguar in February 1964 as Chief Engine Development Engineer. This completed a really formidable team of engineers at Browns Lane, one from which any knowledgeable industry man would expect an outstanding engine to evolve – and it did. Still in charge of all engineering work was Bill Heynes, whose kudos for the XK engine had now been enhanced by the Le Mans racing successes, and by the fantastic reception given to the 'E' Type in 1961. Claude Baily, with whom I had worked on the XK engine in those fire-watching days of 1943 onwards, was his Chief Engine Designer and his Chief Chassis Development Engineer was Bob Knight, who was mainly responsible for the superb standards of refinement and silky performance which every post-war Jaguar had exhibited.

So the 'old firm' was back together again, on an even more exciting project than the original. I was closely involved from 1964 on, but only in an informal capacity at first, as I was still closely controlling Coventry Climax racing activities until the end of 1965. Harry moved in to Jaguar before the first engine was complete, and before a racing sports car to carry it had been designed. The original XJ V-12 engine – not to be confused with the six-cylinder XJs we had in 1945 and 1946 – was a four-cam 5-litre racing unit, with downdraught inlet ports and Lucas fuel injection. It looked splendid and made a lovely noise, but I think we were all disappointed when we found that it was a struggle to extract more than 500 bhp from it, even in racing trim and with a very short stroke. However, having said this, the engine in a suitably streamlined car would probably have been quite good enough to propel a winning combination at Le Mans as it stood – the 7-litre Fords won easily in 1966 and 1967 with around 550 bhp and a lot of Climax-like mid-range torque.

This power output was equivalent to 100 bhp per litre – a figure achieved only once previously by Jaguar when they built the very special racing 3-litre XKs for the prototype 'E' Type in 1960 – which was creditable enough. However, over at Coventry Climax we were now accustomed to seeing more than 130 bhp per litre on Grand Prix designs. Even if one had conceded a little power to aid endurance running, say to 120 bhp per litre, this should have resulted in a 5-litre Jaguar producing around 600 bhp, considerably more than was ever achieved by this four-cam engine. There were, in addition, other puzzles. Not only was there a lack of top-end power, there was also a distinct lack of low-speed and mid-range torque.

An engine lacking both top-end power and mid-range torque was not my idea of a potentially successful racing unit. In a way it reminded me of the elegant engine my old boss, W.O. Bentley, designed for Lagonda; his V-12 looked good and sounded splendid, but it suffered the same characteristics as this Jaguar, namely, a lack of low-down power.

By the time I became involved in the engine's development, its *'raison d'etre'* had changed somewhat from a pure racing to a touring application; consequently, the need for lusty low-speed and mid-range torque had become much more important than top-end power. We also had to make sure that the new engine was quiet and smooth in operation, and that it would satisfy the exhaust emission limitations being imposed all around the world. It was this last requirement which held back production for some time.

Jaguar policy at the time also envisaged two types of engine being evolved from one basic design – hopefully a high-powered sports car version with hemispherical combustion chambers and twin-cam cylinder-heads, and a second version intended to be much cheaper to manufacture, saving money, weight and complication by having single-cam cylinder-heads.

The work that followed to turn the first prototype V-12 engine – a peaky, noisy, unrefined, 500-bhp racing design – into the silkily-smooth production design which powers present-day Jaguars and Daimlers, occupied much of my time from 1965 to 1972. By British standards that is quite a lengthy gestation period, but we must remember that at that time it was probably the most complex quantity-production engine to be built in Britain, and was very costly in many ways. Sir William was faced with this large bill for tooling, and was necessarily determined to delay completion of the new machines until we were absolutely certain of the engine's specification.

In their prototype design for the racing engine, Bill Heynes and Claude Baily had realised that there was very little elbow room inside the vee of a 60-degree V-12 to accommodate efficient inlet ports and passages, and had opted for the alternative of vertical ports, placing carburettors or fuel injection above the cylinder-heads, with mixture being channelled vertically down into the cylinders between the twin camshafts. A lot of development work was carried out on at least three different types of inlet port to this design, and we established that the vertical inlet port was inferior to the more conventional cross-type of port. It is very interesting that the M196 Mercedes-Benz Formula One and sports car engine of 1954/55 also had vertical inlet ports, and also failed to live up to its promise. Although this engine powered many race-winning cars it had somewhat poor torque characteristics, a problem compounded by the use of inlet ports and valves which were too large.

While we went ahead on the 'two types of cylinder-head' philosophy, I could draw on much useful experience gained at Coventry Climax in previous years. I had found very good results at Climax using a completely flat machined cylinder-head in place of the wedge-head which had been a feature of our fire-pump engines. We had converted a 750 cc FWM petrol engine to diesel operation by this means, and when we had studied features in the Heron-headed Rover and Ford V-6 designs we decided to go even further. We decided to standardise this layout with substantially the same cylinder-head for diesel and petrol-engine versions, the main differences being in piston crown recesses and of course substitution of an injector nozzle by a sparking plug. In testing, to our surprise we found that in a petrol engine a flat-head performed rather better than a wedge-head and would accept a compression ratio at least one number higher for a given fuel consumption while giving slightly more power.

To cut down the enormous amount of work needed to assess the qualities and evils of the various alternatives, we decided to make a series of single-cylinder test engines at Jaguar. We could not have reached logical and completely documented conclusions by working on V-12s all the time, and of course costs would have been much higher. A spy in our engine test shops in the mid-1960s might have thought we intended to break into the high-performance motorcycle market, for he would have noticed several water-cooled singles throbbing away, and very few V-12s!

This mass of testing showed the superiority of the finalised flat-head for touring car purposes, and highlighted the poor performance of the hemispherical combustion chambers at low and mid-range speeds. This was very puzzling at first, and contrary to our expectations based on previous engine experience. However, it is now obvious that whereas the XK engine was a long-stroke design in all its most famous forms, the over-square 2.4 and nearly-square 2.8-litre versions were never as satisfactory. My interpretation is that a hemispherical head is better when allied to a long-stroke layout, and not so good for short-stroke layouts unless sheer maximum power is required.

We finally settled on a flat-head, which is unavoidably limited in breathing for high power and high rotating speeds, because of the side-by-side valves. We also made sure that suitable profiling in the piston crowns gave us satisfactory emission characteristics, and with excellent mid-range torque we could still rely on a satisfactory engine performance up to 6,500 rpm. Even this speed was somewhat high for the Borg-Warner automatic transmission we were to use, so one can understand that a four-cam engine revving to 7,000 or 8,000 rpm would have been completely useless in a road car.

The big philosophical obstacle, of course, was in making the change from the XK-based twin-overhead-camshaft cylinder-head to the single-cam head which we had proved to be desirable. If human nature and personal preferences had not been involved we might conceivably have cut down a little on the time and deliberation we went through, but as in any living and active company there were various important views and opinions to be reconciled.

There was no doubt that Bill Heynes and Claude Baily were not really happy to abandon the twin-cam layout, for they knew very well just how much Jaguar owners were devoted to their XK sports cars and the twin-cam saloons. Somehow, a twin-cam head seemed to be an integral part of Jaguar's image, and they thought the V-12 should retain this if at all possible. Sir William took a neutral, and in my view absolutely correct view about this; at the end of the day he wanted an engine of which he could be proud (as he was very proud of his XK engine), one which would perform well, be ultra-refined, and one which by its appearance would maintain an aura of the Jaguar image he had spent so many years building up.

However, there came the day when Sir William called for full reports on both lines of development. He also personally tested cars fitted with both engine types. At a technical board meeting a vote was taken – and the die was cast in favour of the single-cam type of cylinder-heads for production cars. By this time it had also become clear that we could not afford to produce two types of engine – the extra tooling charges could not be justified – and this led to the demise of the twin-cam cylinder-head.

In the midst of all this engineering activity there was much personal and corporate activity to make life as fascinating as ever. Although I moved over from Widdrington Road to Browns Lane in 1966, I kept my position as Technical Director at Coventry Climax, where a major project was the development of a complete family of engines for generator sets for the armed forces and civil defence. This led to much work from the highly versatile design office, culminating in the release of two-cylinder, four-cylinder and V-8-cylinder engines, with diesel-engined equivalents of all of them. There were many parts common to all the range, and the basic four-cylinder unit was a considerably-developed version of the good old FWM fire-pump engine which, very indirectly, had given birth to the successful 1½-litre Grand Prix FWMV.

This might sound a formidable task for such a small office, but my team of designers were well versed in this sort of activity. Nor were they averse to cutting a few corners at times. One good example of the way in which we did unconventional things to save time and money on experimental work was when we cut up a couple of four-cylinder FWM engines, welded them neatly together, and produced a running six-cylinder version without the need to obtain any new castings! The first two-cylinder AFF engine from this family was made in the same way.

In terms of design history and breeding, the V-8 member of the new family, particularly the high-performance CFA, was loosely related to the old FWMV, but was not based on any of its internal dimensions. However, it was an engine that would have been ideally suited for use in a sports car or high-performance touring car if other engines within the group had not already been made available. The gossip-mongers in the motor-sporting press once linked this engine with a high-performance version of the MGB for racing, but it was pure speculation. Just for the record, the CFA V-8 was of 2,496 cc, with a bore and stroke of 80.77 and 60.96 mm, respectively, and it produced more than 200 bhp at 7,000 rpm in flexible sports-car tune. Even more interesting was the fact that the cylinder-heads incorporated single overhead camshafts, in-line vertical valves, and 'bowl in piston' combustion chambers – features which came increasingly into favour at Jaguar as the weeks rolled by!

I was sorry that the CFA could not be put into production, for it was very efficient (more than 200 bhp from only 300 lb weight) and we got as far as fitting one to a works-owned car and made a really exciting machine of it, but in the end the Rover production V-8 got the verdict as the engine for the 'big' MGB.

At the time it was intended to make the Jaguar V-12 the start of a whole new family of Jaguar engines, and I don't think it is breaking any confidences now to say that there could have been a V-8 version on sale if it had measured up to Sir William's exacting standards. There isn't a V-8 XJ saloon, of course, though more than one V-8 prototype was built. Because it was intended to use much of the common V-12 tooling, the V-8 engine had only a 60-degree angle between cylinder banks, and while we all hoped that this would prove satisfactory it was a disappointment to us all. Engineers know that V-8s should have 90 degrees between the banks, but we had hoped that the unpleasant secondary vibrations involved in a 60-degree design could be suppressed by clever engine mounting. But this proved not to be possible and, as with the four-cylinder XK engine of the 1940s, there were unpleasant vibrations in the

We all grow older but still enjoy ourselves. Bill Pacey and I recalling Brooklands days of forty years before when we were reunited with the Pacey-Hassan at Silverstone in 1973

At a Bentley mechanics' reunion, with Les Pennal in the centre and Nobby Clarke to his right

A proud moment in 1971 when I went to Buckingham Palace to collect my OBE. Ethel, Susan and Bill were there to help

structure, felt also through the gear-lever, which we could not tame. There are other possibilities and permutations in the basic V-12 theme, but in view of the somewhat changed world economic environment I doubt if Jaguar will bother to exploit them, at least for some time to come.

It wasn't until March 1971 that the general public first saw the V-12 engine that had occupied our working thoughts for about seven years – and then they saw it in the 'E' Type sports car for which originally it was never intended. Harry Mundy and I were very much involved in the launch preparations, which included extensive written publicity and the making of a film about the V-12. My old friend Raymond Baxter was to narrate the film, and he so helped us over our initial stage fright that what had looked like being an ordeal turned into a very pleasant and interesting experience. There was an unfortunate aside to this film, though. The XJ13, the mid-engined sports-racing car that at one time might have taken Jaguar back to Le Mans, was wheeled out again to show off its performance at MIRA in front of the cameras. On January 20th, 1971, it suffered an accidental tyre deflation at more than 150 mph, and rolled over several times, damaging itself severely, although I'm happy to say that neither its driver, chief tester Norman Dewis, nor the car were irretrievably harmed. The car was lovingly restored by Phil Weaver and his old competitions department mechanics, and made several stirring demonstration laps at Silverstone before the 1973 British Grand Prix race. The nearest thing we get to a racing Jaguar V-12 at Silverstone these days is the high-speed XJ12 fire tender which now operates at the circuit, but even with all that light water on board it is apparently very competitive when it follows the field at the start of some saloon car races, and at times I reckon its driver is sorely tempted to join in!

While we were reaching the climax of our efforts on the new V-12 engine, in the summer of 1969, Bill Heynes retired. He had presided over the design and development of many outstanding Jaguar products since 1935, but from the mid-1960s he had delegated many of his responsibilities on engine matters to me. Over the years he had become very close to Sir William, and had finally been appointed Vice-Chairman (Engineering). When it became known that he was to leave the company at the end of July there was lively speculation among the staff as to Sir William's intentions. I need hardly say that I was both surprised and delighted to be told that both Bob Knight and myself were to be invited to join the Board of Directors from August 1st. Bob was to oversee all vehicle engineering, and I was to look after group power units development.

But here was a rather odd situation. Bill Heynes was retiring in his 66th year, and it had become general policy within Jaguar that one would retire at 65. Yet here I was, newly appointed to the board, and already past my 64th birthday. Was I really intended only to have nine months in this new post before leaving full-time employment in industry for good?

Since I was rather looking forward to the new challenge of general management, I thought I should check it out with my superiors. Accordingly, I had a few words with 'Lofty' England, an old Brooklands friend who was by then Joint Managing Director of the company. Could he confirm that I should retire in April the following year? Should I begin to sort out the succession to my job, and all the reshuffling which would follow it? Should I begin to train people to take over from me in the various design, standards and policy-making committees I now attended within the recently formed British Leyland Motor Corporation?

'Lofty' came back to me, saying that he had had words with Sir William, and he understood that Sir William wanted to apply the '65th birthday' ruling to everyone. I was a little disappointed, but not sad, and I sent a memo to Sir William detailing the changes I thought should be made before I left, so that there should be a smooth changeover in the ensuing months. I then decided to sort out my private life, but in the meantime went off to the opening days of the 1969 Earls Court Motor Show to enjoy myself.

However, a misunderstanding must have arisen, because on the opening day I had two telephone calls from Sir William to the Jaguar stand. The details are unimportant, but the substance was that he certainly didn't intend to let me go so soon, and would I arrange to see him as soon as I got back to Coventry! This I did, where Sir William made it clear that he would like me to stay on until the V-12 engine was in production. We decided that this would be achieved in 1971 at the earliest, and probably in 1972 if all model ranges were considered, so I was asked to continue, past normal retiring age, until my 67th birthday in 1972.

Apart from the continual uphill struggle to get the V-12 ready – an 'uphill struggle' only because of the way the North American exhaust emission regulations kept changing and making us undertake further proving programmes – I was fully occupied at Browns Lane for a time. 1969 and 1970 were years when new regulations poured out of the US safety and environmental offices in Washington like a paper torrent, though a year or so later several proposed standards were withdrawn, several were modified, and several more were put on ice. Yet because of the time-scales involved, whenever we saw a new set of limits – especially the terribly difficult standards proposed by the Clean Air Act (the notorious 'Muskie Bill' whose requirements several years later are still being altered and pummelled into an acceptable shape to allow the big US car makers to stay in business!) – it meant that all the various alternatives had to be designed and tested so that we could be in a position to meet any of the regulations should they become law.

Another aggravation – only professionally aggravating, because this is all part of an executive's job – was a decision to bring forward the introduction of a V-12 'E' Type and to defer introduction of a V-12 saloon. In terms of engine accessories, controls and test procedures to be completed, there were many differences between the two versions of the engine, and with a load of priorities to be re-allocated this all added to the pressure. I was reminded of the passage in the book of Common Prayer, which says something like: "We have left undone those things we ought to have done, and we have done those things which we ought not to have done . . ." because that is what it seemed like for a time.

On top of all this was my increasing involvement with other member companies of British Leyland. The merger, which took place in 1968, linked Jaguar and BMC with Leyland-Triumph and Rover, and almost by definition it meant that there was a proliferation of engine and vehicle designs that would have to be rationalised in the coming years. The wrinkle that seemed to amuse the motoring press was that after years of saying that British car makers were too reluctant to introduce new engine designs, the newly-formed British Leyland Motor Corporation was flush with them. Certainly in terms of vee-formation units this was true, for apart from the Rover V-8, which had already been announced, there were the Triumph Stag V-8 and Jaguar's V-12 and proposed V-8, all still under wraps but active.

One of the first things Lord Stokes, as Chief Executive, decided was that there should be a series of design co-ordination groups, and I was privileged to be asked to chair the group concentrating on future British Leyland engine policy. I was also asked to run a joint anti-pollution policy committee for British Leyland, for whom the group emission test facility was located at Browns Lane.

Membership of these committees meant that not only was I concerned with Jaguar power units, but I had to co-ordinate the engineers of companies which in the past were in active commercial opposition; it was very satisfying that without exception they co-operated very successfully. These committees did, and still do, work well, and were extremely useful in providing a technical forum where everyone's difficulties could be freely discussed and useful advice given across the table. I can illustrate this by recalling that Spen King was asked to formulate proposals for a high-performance version of the overhead-camshaft Triumph Dolomite engine; in due course he talked to Harry Mundy and myself about a four-valve layout. Both Harry and myself were completely convinced of the worth of a four-valve set-up, and we were able to confirm Spen's ideas that the Dolomite Sprint engine should be so equipped. We were also able to pass on results of tests on port shapes, valve sizes and included angles which we had collected over the years. At the same time, I would not like to detract from Spen's fine efforts to make a production proposition out of it; we had nothing to do with the actual design, nor the particularly clever way in which all sixteen valves are worked from a single camshaft, nor with the refinement and docility built in to the production car. It's nice to see how 'tuneable' the engine is, for Ralph Broad's and other entrants' Sprints have since performed extremely well in Group 1 saloon car racing.

Even though we had withdrawn from motor racing years earlier, I couldn't keep away from the scene, nor from discussing progress with other designers. In the closing years of my time at Jaguar, I often talked to Keith Duckworth about his Cosworth Ford DFVs and their little problems. In many ways the DFV was a logical development of the work we had carried out at Coventry

Climax. The four-valve layout was broadly the same, though the included valve angles were reduced to 32 degrees; our racing experience had indicated that we should continue to reduce the included angles – the FPF angle was 66 degrees, the FWMV V-8 was 60 degrees (two-valve and four-valve), and the unraced flat-16 FWMW was 48 degrees – and any engine we might have raced after the flat-16 would certainly have approached Cosworth's figure. I notice also that Keith's engine was designed with a flat-crank from the word 'go', which eliminated any of the exhaust-system conundrums we had had to solve a few years earlier. When the DFV was first released its specific output was marginally down on that of the four-valve Coventry Climax V-8, but it has since risen to much higher levels. Keith is a ruthlessly logical individual who would design every little component from first principles if he could only find the time, but he is also refreshingly honest about the inevitable problems which turn up and which he may not immediately understand. The vibration problems which led to breakages in 1969 and 1970 nearly drove him scatty, and we had several talks about similar things that had occurred on our own engines. Of course, engineers always enjoy talking and comparing notes!

Meanwhile, at Browns Lane, I had to arrange for my successor to be appointed, and once again I had little hesitation in recommending Harry Mundy for the job. Harry had become Chief Engine Designer on Claude Baily's retirement a few years earlier, and we were all quite sure he could carry on with V-12 and other developments in his usual inimitable manner. Harry became Deputy Chief Engineer a year before I was due to go, and took over from me as Chief Engineer in 1972, which was the fulfilment of my original thinking when he left *The Autocar* much earlier.

During my last year it was both satisfying and rather amusing to observe the way in which I had been able to delegate my various duties. I wanted to make sure that on my 67th birthday I could step out of my office leaving an absolutely clean desk, and not be missed. Metaphorically, I tried to arrange things so that I could have broken my neck at any time in the last few months and nobody would have been inconvenienced – I nearly said so that no-one would have noticed, but I am a little too vain for that! By the end of 1971 I was actually looking forward to having some spare time, and I honestly thought that at Coventry I had come to the climax of a long career among fine engines, fast cars and motor racing. It was all very attractive in prospect, but fate still had another card to play. Before I could take part in the promised presentation ceremony, and leave Browns Lane for the last time, I had a phone call – and at 67 years old I was pitched back into motor racing once again!

Jaguar V-12 engine development

Before Jaguar withdrew from racing at the end of 1956 there had been thoughts of replacing the 'D' Type competition car with a much more powerful prototype.

Project studies for a new V-12 engine started then, and carried on at low priority for a number of years. The original V-12 layout was based on a couple of 'D' Type racing cylinder-heads, and the engine's capacity could have been anything from about 5-litres to nearly 8-litres if the 3.8-litre bore and stroke had been chosen. When the project was seriously reactivated in the early 1960s the main competition was from Ford of America and Ferrari, and it was decided that a 5-litre engine size was desirable to slot the engine in neatly under the capacity limit then impending.

Original design was directed by Bill Heynes, whose Chief Designer was still Claude Baily. The first engine ran in August 1964, and the mid-engined XJ13 it was to power was completed in 1965, by which time Ford had pre-empted the game by fitting a big 'soft' 7-litre in their Le Mans prototypes! The racing project then languished due to lack of commitment and to the developing interest in the engine as a production unit (it would have been wrong to show off the 'racer' at that time, even as an obsolete project, if the engine could not speedily be sold to the public). So it was not until March 1967 that the car was first run at MIRA, when everything went well, David Hobbs lapping the tight banked circuit at 161 mph. Later the same year it was tested at Silverstone, but then sidelined because the car was now obsolete in terms of tyre widths and suspension behaviour. For the announcement of the production V-12 engine in 1971, the old car was brought out of retirement to perform once again at MIRA for cameramen making a film about the engine, when as already recorded it was badly damaged in a very high-speed accident. After being painstakingly rebuilt in its original form, and exhibited at the British Grand Prix meeting at Silverstone, in 1973, the XJ13 car and its racing engine is now preserved by Jaguar as a fascinating museum piece.

The original racing engine had a bore and stroke of 87 × 70 mm, with a 10.4 to 1 compression ratio, twin overhead camshafts with vertical inlet ports and slip-fit dry cylinder liners. Swept volume was 4,994 cc, maximum power 502 bhp at 7,600 rpm, and maximum bmep 191 psi at 6,300 rpm. It weighed 648 lb, or 144 lb more than the production single-cam unit.

Even in racing form the power output was no better than 100 bhp/litre, which did not satisfy Hassan and Mundy, whose FPE eight-cylinder engines had achieved this more than a decade earlier. The deficiency was due to poor top-end breathing, and insoluble without drastic revision of the inlet port arrangements; nevertheless, the limitations of space in an XJ bodyshell meant that horizontal inlet ports were out of the question on the outside of the engine, and there was little enough space inside the 'vee' for any type of efficient manifolding.

Hassan and Mundy then embarked on a development programme using single-cylinder test engines, but with great wariness. A comment in Harry Mundy's I.Mech.E. paper states: ".... it is dangerous to read across from single-cylinder work to full-scale engines as the resonant frequencies of inlet and exhaust systems at full throttle can give surprisingly misleading results; but used intelligently a lot of useful information can be obtained". One brief from management was to 'tame' the engine to give about 330 bhp (gross), to provide the *adequate mid-range torque which the racing version lacked, to keep the weight within 80 lb of the old XK unit, and to make sure the engine would fit into existing engine bays of models either in production or under active development. Effectively this meant that the XJ6 bodyshell was tailored to a provisional profile agreed as early as 1964, while the same unit had to be squeezed into the complex spaceframe of the 'E' Type. There were, and are, other versions of the basic engine concept which may still not be described. However, as Walter Hassan has explained, a V-8 version was seriously considered for a time, which retained the 60-degree angle between cylinder banks, and could have had a capacity between 3-litres and 4-litres depending on marketing requirements.*

Much soul-searching went into the type of cylinder-head which should be chosen for the production engine. Hassan and Mundy had amassed a great deal of successful practical experience with a single-cam head at Coventry Climax, yet the twin-cam layout was almost 'Holy writ' at Jaguar and much favoured by other top executives. No fewer than four types of single-cylinder test engine were built with single-overhead-camshafts, vertical valves and flat decks - the combustion chambers being formed completely in the piston crowns - to assess the differing arrangements of ports and sparking plug position. This arrangement had proved to be superior to the original wedge-head layout at Coventry Climax, which was never actively considered for the V-12. Mundy's paper to the I.Mech.E., and Hassan's to the American S.A.E., make it clear how much basic research was carried out on single-cylinder test engines, at the end of which it became clear that this layout could be made superior to the original twin-camshaft layout at all speeds up to 5,000 rpm; it also became clear that the vertical inlet port was superior at all speeds likely to be experienced in production engines, though for sustained speeds above 6,000 rpm a horizontal port exhibited better filling. Breathing comparisons between the racing and production engines would be completely meaningless because of the different port diameters, valve sizes and timing diagrams used; however, there was little doubt that the single-cam head could produce the power required very easily indeed.

It was not until the decision to go for this arrangement had been taken that construction of the first production-type 12-cylinder engine was attempted. Once building began, other advantages conferred by the simpler design became clear; the cylinder-heads were each 26 lb lighter, and there was a reduction in overall width without a corresponding height penalty. The cylinder block was much revised, eventually appearing with a Coventry Climax-type of open-deck construction, including wet cylinder liners. The bore size was increased to 90 mm, and swept volume to 5,343 cc, to provide acceptable torque.

Development, however, was protracted, and the XJ6 range of saloons was released in the autumn of 1968 with six-cylinder engines only, and the promise that 'new engines would be available within two years'. This did not transpire, however, and it was two-and-a-half years before the V-12 engine was shown publicly, and then only in the 'E' Type sports car. It was not offered in the XJ12 and Daimler Double-Six saloons until June 1972, nearly four years after it was first mentioned, and nearly ten years after the decision had been taken to fit a V-12 into production cars, although in the early part of 1968, when Jaguar had held informal briefings with some of the motoring press, it had still been hoped to have the engine in production by announcement day for the new saloons later that year. Much of the delay may be attributed to the exigencies of complying with progressively more restrictive United States exhaust emission regulations, which nevertheless are met by the V-12 engine with the minimum of 'hang-on' aids to clean combustion.

When first announced in the 'E' Type in 1971 the engine was credited with a maximum power output of 272 bhp (DIN) at 5,850 rpm, and maximum torque of 304 lb/ft at 3,600 rpm. This compared with 171 bhp (DIN) at 4,500 rpm and 230 lb/ft torque at 2,500 rpm for the 4.2-litre XK engine in the same car.

At the time of writing the XJ engine is the only V-12 in large-volume production anywhere in the world, and since 1945 the only other V-12s made in any reasonable quantity have been from Ferrari and Lamborghini. Before 1939 there were several V-12s in small-volume production, all except the side-valve designs from Ford and Lincoln being built to power luxurious prestige cars. From Britain there were the Phantom III Rolls-Royce and the W.O. Bentley-designed Lagonda, along with the monstrous Daimler Double Sixes, while Marmon in America and Hispano-Suiza were similarly powered. However, more V-12 Jaguar engines were to be made in the first three years of production at Coventry than the sum total of all other V-12 engines previously produced for use in passenger cars.

10
And they call it retirement!

For 1972 I looked forward to the prospect of pottering about in my boats, one of which was kept near Leamington and the other in Cornwall, where I also had a house, and I was savouring the thought of some leisurely fishing. I would be able to call all of my time my own, and begin to catch up on the many pleasures denied me by a busy working life. However, this was not to be. One day, quite out of the blue, I had a telephone call from Louis Stanley of BRM, saying, in effect: "Now that you have no other business ties, how would you like to act as a consultant to BRM on V-12 engine developments for our Formula One car?"

This was not the first time I had been approached by the Owen Organisation, of which BRM was then a part. Way back in 1952, Sir Alfred Owen had asked me to leave Coventry Climax to join him, to go out to Australia, and take a hand in starting up a fork-lift truck enterprise for him. I turned that down flat; I had only been with Coventry Climax for a couple of years, and was already very well established and happy there. The work involved would not have been new to me, in fact it would have been absolutely similar to that at Coventry, and since I didn't really fancy emigrating at my time of life I was not tempted to go.

Later still, in the 1960s, Louis Stanley had approached me several times while I was at Coventry Climax, during the time we were still actively involved in motor racing, and invited me to join BRM to take over responsibility for racing engine design at Bourne. Now everyone knows that Coventry Climax and BRM were motor-racing rivals at the time, and although 'poaching' (the term used in the motor industry for encouraging people to move from one firm to another making very similar products) was quite cheerfully accepted in the industry, I felt great loyalty to Leonard Lee and decided not to move. I thanked Mr Stanley for his offer and said that I could not consider anything like this until I retired. Obviously he had filed this remark away in his memory, and came back to me as soon as he could.

In the meantime, I had been instrumental in encouraging Peter Windsor-Smith to move to Bourne to take over engine design in 1972. Peter had been my

Chief Designer during the final successful years at Coventry Climax, and following the closure of racing activities had moved around in the British Leyland group. At one time he had been Chief Engineer (Bus Design) at Daimler, but later he moved up to Leyland Motors in Lancashire.

Through the grapevine I came to hear that Aubrey Woods was leaving BRM, and I knew that the vacant job would be absolutely ideal for Peter if he still wanted to design racing engines. I contacted him in Lancashire, and eventually he went back to his old love; it must have made quite a change from buses and diesel engines! With Peter already at Bourne, I was very attracted to the idea of part-time work in motor racing once again, and as soon as possible I started a series of visits to Bourne. If only Harry Mundy had been with us it would have been a full reunion of our design team of the 1950s and 1960s, but as I saw a lot of Harry socially anyway we were still swapping ideas and experiences fairly regularly.

Earlier I said that at Brooklands in the 1930s it was possible to get special components made in an astonishingly short time, but that by the time similar things were needed for development purposes at Coventry Climax it was by no means as easy. I soon found, at BRM, that the problems of administration and getting delivery of a small quantity of special high-performance parts had become even more difficult. I am not trying to make excuses, either for myself or for BRM, when I say that changes agreed in 1972 were still not fitted to the Grand Prix cars at the start of the 1975 season. The frustrations over the slow supply of essential components seems to be an ever-increasing problem for racing teams seeking to improve their standard of competitiveness. It is all too easy to criticise a poor performance from the security of the grandstand, and without knowledge of all the facts.

Some journalists have said that the BRM V-12 engine is a very old design, and have even suggested that its ancestry dates back to the first V-8 BRM of 1961. Of course there is a recognisable pattern of development throughout – for instance the 1½-litre V-8 became the 2-litre for the Tasman series, the V-12 at first was really based on the Tasman engine's dimensions and cylinder-head, and so on – but the latest engine really has little in common with the 1961 design. The cylinder-heads, the heart and lungs of any engine, are completely different. They have four-valves-per-cylinder, of course, like our final successful Coventry Climax designs, though the included valve angles and porting details have been revised several times. Twelve cylinders are the maximum number allowed by existing Formula One regulations, and in terms of maximum piston area and potential maximum rpm this layout must still be theoretically superior to that of a V-8. The problem, of course, with a V-12 is that there are so many more opportunities for frictional and other losses to occur, and the maintenance and adjustment problems are that much more severe.

Although we can never really know what the ultimate in specific power output will be, it may be quite some time before a 3-litre racing engine reaches much more than 500 bhp. That sort of figure is well within the bounds of possibility, of course, but it must come from an ideally designed combustion chamber, something which seems to have been achieved already. Optimising everything from fuel injection to porting, and camshaft timing to breathing will all help, but one of my strongest theories is for cutting down on internal losses. Keith Duckworth was very clever when he designed the DFV; he really set out to keep down churning losses in the crankcase of that engine and it has really paid-off.

If only we could get away with materials which would act like asbestos for liners, piston tops and heads we could retain more heat, which must be a good thing; after all, heat is power. A nice thought, but hardly practical! Keeping everything nice and shiny in the combustion area is also a good thing (radiated heat is directed back into the nucleus of the flame) but carbon build-up soon puts a stop to that. Way back in Brooklands days it was it was well-known, for example, that a racing motorcycle with a highly polished piston top and cylinder-head would be measurably faster over the first half-lap than later, when the polish had been lost.

I suggested earlier that combustion chamber design has virtually reached the ideal. All racing engine designers now agree that four-valves-per-cylinder are essential, though there is still some disagreement over the angles between them. Our first four-valve head at Coventry Climax had 60 degrees between the stems, the same as in the two-valver in which this seemed to be the optimum, but we found with the four-valve head that we had a slow-burning mixture problem which we eventually put down to the rather severe pent-roof profile of the pistons. If we had had time to reduce these angles we would certainly have improved matters even further. Designers nowadays seem to have gone to the other extreme, and I have seen designs with included angles of not more than 12 degrees. I don't approve of this because to get the compression ratio right one needs a piston top with turrets rather like a miniature Windsor Castle; not only do these get hot, but they cut down turbulence, which we have always found to be very useful.

However, getting the power is not the only problem with racing car design these days; equally important is finding the time to do any development. With a season that starts early in January in Argentina and ends at the end of October in the United States, with testing to be done outside Great Britain in the winter months and at least three cars to be kept running, it's hardly surprising that pure development sometimes takes a back seat to chassis builds, rebuilds and power tests.

Now, at 70 years of age, I too find that life seems as busy as ever, after something like fifty-five years of close connection with motor racing and the motor industry. Although I am not as mobile as I used to be, I seem to do more travelling on business nowadays than during my last few years at Jaguar, and I am very happy to say that I have not yet decided to stop working altogether, as I still find every new aspect of engineering, particularly in engine design, of

The reason for my deferred retirement. The Jaguar V-12 engine, a magnificent piece of engineering which Sir William Lyons asked me to see through to production before taking my leave. This is the engine in its 1975 form, installed in Jaguar's prestige model, the XJ-S

I have always liked messing about in boats. This is my ex-naval 25-foot cutter; one day I will get it finished

Retirement house—well, one day perhaps. This is our second home, near Falmouth, where I keep a boat quite literally at the bottom of the garden

great interest. Although I admit to getting a bit lazy, I still can't foresee a time when I might want to put my feet up and drop out of engineering for good.

If I did, I am fairly certain that my time would be called upon in many other nice ways. Various clubs have asked me to talk to them about my experiences, and they seem to be as interested in my work at Bentley and in the 1930s as in the later racing years. I am always happy to oblige these people, although strangely enough I am not really interested in what I would call obsolete machinery. In fact I don't think I would cross the road now to look at an old Bentley. However, if people are keen to hear about the old days - and it is people who regenerate the interest in me - I am happy to talk about them. But make no mistake about this, the people who look back on the 1920s as a Golden Age in engineering are wrong. Engineering advances didn't just stop at the end of 1930 as the Vintage enthusiasts would have us believe; we're much more concerned with detail now than ever we were.

I have come a long way since I was a ten bob a week shop boy at Bentley. Most of the time it has been great fun and I have thoroughly enjoyed my life. I am very sad that my sons and their sons will never find quite the same excitement in life, nor perhaps as many new things to experience. Somehow motoring, real motoring, isn't as easy to find, and it's certainly not as much fun. For my part I am content. I have experienced successes and great happiness, and even some failures. I would not willingly have changed it.

Appendix 1

Coventry Climax Grand Prix record 1957-1966

The first twin-cam FPF Climax engine made its Grand Prix debut at Monaco in May, 1957. In 1,960 cc guise it would have propelled Jack Brabham's Cooper to third place if the fuel pump hadn't fallen off just ten miles from the end; the gallant Brabham pushed home into sixth place! Eight months later Stirling Moss won the 1958 Argentine Grand Prix in his 1,960 cc FPF-engined Cooper-Climax and the success story was under way.

Coventry Climax officially retired from Grand Prix racing at the end of 1965, though Mike Spence's Lotus won the South African GP in 1966. Between 1958 and 1966 Coventry Climax engines powered cars to 96 Grand Prix wins, of which 40 were World Championship events.

Of these 96 victories, Lotus cars accounted for 59, Cooper 27, Brabham six, Lola two, and Ferguson (the P99 four-wheel-drive car) and BRM (a 1961 $1\frac{1}{2}$-litre) one each. The 40 World Championship victories were shared by Lotus (24), Cooper (14) and Brabham (2).

Up to that time no single make of engine had powered so many Grand Prix winners, the next challenger being Alfa Romeo with 82 wins and Ferrari with 81. The ubiquitous Cosworth-Ford DFV engine and its derivatives have since passed this total in a life extending over eight years.

The most successful World Championship driver of a Coventry Climax-powered car was the late Jim Clark, whose Lotuses gave him 19 wins between 1962 and 1965. Behind him were Jack Brabham and Stirling Moss, each with seven wins.

The following table details the victories, the events and the engines used:

Event and circuit		Driver and Car	Engine	
1958				
Argentine GP	Buenos Aires	Stirling Moss (Cooper)	1,960cc	FPF
Monaco GP	Monte Carlo	Maurice Trintignant (Cooper)	2,015cc	FPF

 – also 1 second place and 2 thirds during the year.

APPENDIX 1

Event and circuit		Driver and Car	Engine	
1959				
Monaco GP	Monte Carlo	Jack Brabham (Cooper)	2,495cc	FPF
British GP	Aintree	Jack Brabham (Cooper)	2,495cc	FPF
Portuguese GP	Lisbon	Stirling Moss (Cooper)	2,495cc	FPF
Italian GP	Monza	Stirling Moss (Cooper)	2,495cc	FPF
United States GP	Sebring	Bruce McLaren (Cooper)	2,495cc	FPF

– also 3 second places and 4 thirds during the year.

1960				
Argentine GP	Buenos Aires	Bruce McLaren (Cooper)	2,495cc	FPF
Monaco GP	Monte Carlo	Stirling Moss (Lotus)	2,495cc	FPF
Dutch GP	Zandvoort	Jack Brabham (Cooper)	2,495cc	FPF
Belgian GP	Spa	Jack Brabham (Cooper)	2,495cc	FPF
French GP	Reims	Jack Brabham (Cooper)	2,495cc	FPF
Portuguese GP	Oporto	Jack Brabham (Cooper)	2,495cc	FPF
British GP	Silverstone	Jack Brabham (Cooper)	2,495cc	FPF
United States GP	Riverside	Stirling Moss (Lotus)	2,495cc	FPF

– also 7 second places and 6 thirds during the year. Coventry Climax-engined cars won every Grand Prix in which they were entered.

1961				
Monaco GP	Monte Carlo	Stirling Moss (Lotus)	1,498cc	FPF
German GP	Nurburgring	Stirling Moss (Lotus)	1,498cc	FPF
United States GP	Watkins Glen	Innes Ireland (Lotus)	1,498cc	FPF

– also 4 third places during the year.

1962				
Monaco GP	Monte Carlo	Bruce McLaren (Cooper)	1,495cc	FWMV V8
Belgian GP	Spa	Jim Clark (Lotus)	1,495cc	FWMV V8
British GP	Aintree	Jim Clark (Lotus)	1,495cc	FWMV V8
United States GP	Watkins Glen	Jim Clark (Lotus)	1,495cc	FWMV V8

– also 5 second places and 4 thirds during the year.

1963				
Belgian GP	Spa	Jim Clark (Lotus)	1,495cc	FWMV V8
Dutch GP	Zandvoort	Jim Clark (Lotus)	1,495cc	FWMV V8
French GP	Reims	Jim Clark (Lotus)	1,495cc	FWMV V8
British GP	Silverstone	Jim Clark (Lotus)	1,495cc	FWMV V8
Italian GP	Monza	Jim Clark (Lotus)	1,495cc	FWMV V8
Mexican GP	Mexico City	Jim Clark (Lotus)	1,495cc	FWMV V8
South African GP	East London	Jim Clark (Lotus)	1,495cc	FWMV V8

– also 6 second places and 4 thirds during the year (all Mark III units).

APPENDIX 1

Event and circuit		Driver and Car	Engine	
1964				
Dutch GP	Zandvoort	Jim Clark (Lotus)	1,497cc	FWMV V8
Belgian GP	Spa	Jim Clark (Lotus)	1,497cc	FWMV V8
French GP	Rouen	Dan Gurney (Brabham)	1,497cc	FWMV V8
British GP	Brands Hatch	Jim Clark (Lotus)	1,497cc	FWMV V8
Mexican GP	Mexico City	Dan Gurney (Brabham)	1,497cc	FWMV V8

– also 2 second places and 5 thirds during the year (all Mark IV units).

1965

South African GP	East London	Jim Clark (Lotus)	1,497cc	FWMV V8
Belgian GP	Spa	Jim Clark (Lotus)	1,497cc	FWMV V8
French GP	Clermont-Ferrand	Jim Clark (Lotus)	1,497cc	FWMV V8
British GP	Silverstone	Jim Clark (Lotus)	1,497cc	FWMV V8
Dutch GP	Zandvoort	Jim Clark (Lotus)	1,497cc	FWMV V8
German GP	Nurburgring	Jim Clark (Lotus)	1,497cc	FWMV V8

– also 2 second places and 6 thirds during the year. All Clark's wins except in South Africa were with the four-valve Mark VI unit.

1966

One third place was gained by Jim Clark's 1,974 cc FWMV-engined Lotus 33.

Appendix 2

Coventry Climax FWMV development record – 1961-1966

Year	Mark	Dimensions	Carburation	Max. power, bmep and torque
				bhp/rpm lb in^2 lb ft/rpm
1961/62	I	63 × 60mm, 1495cc, CR 10.4:1, inlet valves 1.3in, exhaust pipes 1.25in	4 × Type 38DCNL Webers	181/8500 196 118/7500
1962	II	As Mark I except inlet valves 1.35in	,,	186/8500 198 119/7500
1963	III	67.94 × 51.56mm, 1495cc, CR 11:1, inlet valves 1.35in	Lucas fuel injection	195/9500 195 118/8000
1964	IV	72.39 × 45.47 mm, 1497cc, CR 12:1, inlet valves 1.35in, exhaust pipes 1.37 in	,,	200/9750 193 117/8000
1964	V	As Mark IV except inlet valves 1.4in (one engine only)	,,	203/9750 191 115/8000
1965	VI	As Mark IV except two inlet valves 1.04in (one engine only)	,,	212/10300 197 119/8000
1965	VII	As Mark VI except two inlet valves 1.107in (one engine only)	,,	213/10500 189 115/7000 to 9000

APPENDIX 2

Year	Mark	Dimensions	Carburation	Max. power, bmep and torque
1966	IX	72.39 × 60mm, 1974cc, inlet valves 1.35in (one engine only)	Lucas fuel injection	240/8800 194 155/7500
1966	X	As Mark IX except inlet valves 1.4in	,,	244/8900 198 158/7500

For comparison purposes:

| 1965 | FWMW Flat-16 | 54.1 × 40.64mm, 1495cc | ,, | 209/12000   165   100/9000 |

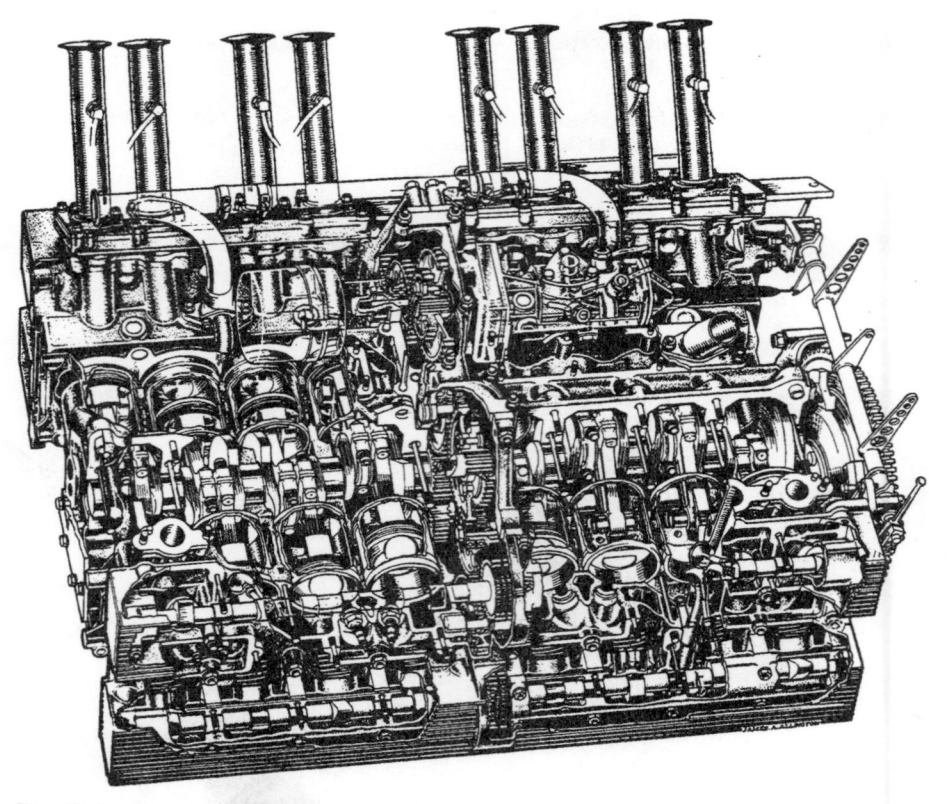

The FWMW in cutaway. It looks complicated but is really four four-cylinder engines joined together. We had plans for four-valve heads, meaning 64 valves in all! Power take-off was through the central gear train and beneath the crankshaft line